NIST Technical Note 1744

Annex 47 Report 2:
Commissioning Tools for Existing and
Low Energy Buildings

Christian Neumann
Dirk Jacob
Fraunhofer Institute for Solar Energy Systems
Henk Peitsman
Netherlands Organization for Applied Scientific Research
Harunori Yoshida
Kyoto University
Hideki Yuzawa
Nikken Sekkei Research Institute
Takeshi Watanabe
NTT Facilities
Katsuhiro Kamitani
Tonets Corporation
Hiroshige Kikuchi
Hitachi Plant Technologies
Oliver Baumann
Ebert and Baumann Consulting Engineers
Daniel Choinière
Natural Resources Canada
Canada Centre for Mineral and Energy Technology
Energy Technology Centre
Natascha Milesi Ferretti
National Institute of Standards and Technology
Engineering Laboratory

http://dx.doi.org/10.6028/NIST.TN.1744

September 2012

U.S. Department of Commerce
Rebecca Blank, Acting Secretary

National Institute of Standards and Technology
Patrick D. Gallagher, Under Secretary of Commerce for Standards and Technology and Director

CITATION

Neumann, C. et al. Commissioning Tools for Existing and Low Energy Buildings, a Report of Cost-Effective Commissioning of Existing and Low Energy Buildings. NIST Technical Note 1744. May 2012.

This publication was simultaneously published by Natural Resources Canada as Neumann, C. et al. Commissioning Tools for Existing and Low Energy Buildings, a Report of Cost-Effective Commissioning of Existing and Low Energy Buildings, May 2012.

Copies of this report may be obtained from the Annex 47 web site at: http://www.iea-annex47.org , the NIST website at: http://www.nist.gov, or from the IEA/ECBCS Bookshop at: www.ecbcs.org.

International Energy Agency
Energy Conservation in Buildings and Community Systems Programme

Preface

International Energy Agency

The International Energy Agency (IEA) was established in 1974 within the framework of the Organization for Economic Co-operation and Development (OECD) to implement an international energy program. A basic aim of the IEA is to foster co-operation among the twenty-eight IEA participating countries and to increase energy security through energy conservation, development of alternative energy sources and energy research, development and demonstration (RD&D).

Energy Conservation in Buildings and Community Systems

The IEA coordinates research and development in a number of areas related to energy. The mission of one of those areas, the Energy Conservation for Building and Community Systems Program (ECBCS) is to develop and facilitate the integration of technologies and processes for energy efficiency and conservation into healthy, low emission, and sustainable buildings and communities, through innovation and research.

The research and development strategies of the ECBCS Program are derived from research drivers, national programs within IEA countries, and the IEA Future Building Forum Think Tank Workshop, held in March 2007. The R&D strategies represent a collective input of the Executive Committee members to exploit technological opportunities to save energy in the buildings sector, and to remove technical obstacles to market penetration of new energy conservation technologies. The R&D strategies apply to residential, commercial, office buildings and community systems, and will impact the building industry in three focus areas of R&D activities:

- Dissemination;
- Decision-making; and
- Building products and systems.

Participating countries in ECBCS: Australia, Austria, Belgium, Canada, P.R. China, Czech Republic, Denmark, Finland, France, Germany, Greece, Italy, Japan, Republic of Korea, the Netherlands, New Zealand, Norway, Poland, Portugal, Spain, Sweden, Switzerland, Turkey, United Kingdom and the United States of America.

The Executive Committee

Overall control of the program is maintained by an Executive Committee, which not only monitors existing projects but also identifies new areas where collaborative effort may be beneficial. To date the following projects have been initiated by the executive committee on

Energy Conservation in Buildings and Community Systems (completed projects are identified by (*)):

Annex 1:   Load Energy Determination of Buildings (*)

Annex 2:   Ekistics and Advanced Community Energy Systems (*)

Annex 3:   Energy Conservation in Residential Buildings (*)

Annex 4:   Glasgow Commercial Building Monitoring (*)

Annex 5:   Air Infiltration and Ventilation Centre

Annex 6:   Energy Systems and Design of Communities (*)

Annex 7:   Local Government Energy Planning (*)

Annex 8:   Inhabitants Behavior with Regard to Ventilation (*)

Annex 9:   Minimum Ventilation Rates (*)

Annex 10:   Building HVAC System Simulation (*)

Annex 11:   Energy Auditing (*)

Annex 12:   Windows and Fenestration (*)

Annex 13:   Energy Management in Hospitals (*)

Annex 14:   Condensation and Energy (*)

Annex 15:   Energy Efficiency in Schools (*)

Annex 16:   BEMS 1- User Interfaces and System Integration (*)

Annex 17:   BEMS 2- Evaluation and Emulation Techniques (*)

Annex 18:   Demand Controlled Ventilation Systems (*)

Annex 19:   Low Slope Roof Systems (*)

Annex 20:   Air Flow Patterns within Buildings (*)

Annex 21:   Thermal Modelling (*)

Annex 22:   Energy Efficient Communities (*)

Annex 23:   Multi Zone Air Flow Modelling (COMIS) (*)

Annex 24:   Heat, Air and Moisture Transfer in Envelopes (*)

Annex 25:   Real time HVAC Simulation (*)

Annex 26:   Energy Efficient Ventilation of Large Enclosures (*)

Annex 27:   Evaluation and Demonstration of Domestic Ventilation Systems (*)

Annex 28:   Low Energy Cooling Systems (*)

Annex 47

The objectives of Annex 47 were to enable the effective commissioning of existing and future buildings in order to improve their operating performance and to advance the state-of-the-art of building commissioning by:

Extending previously developed methods and tools to address advanced systems and low energy buildings, utilizing design data and the buildings' own systems in commissioning;

Automating the commissioning process to the extent practicable;

Developing methodologies and tools to improve operation of buildings in use, including identifying the best energy saving opportunities in HVAC system renovations; and

Quantifying and improving the costs and benefits of commissioning, including the persistence of benefits and the role of automated tools in improving persistence and reducing costs without sacrificing other important commissioning considerations.

To accomplish these objectives Annex 47 has conducted research and development in the framework of the following three areas:

Initial Commissioning of Advanced and Low Energy Building Systems

This area addressed what can be done for (the design of) future buildings to enable cost-effective commissioning. The focus was set on the concept, design, construction, acceptance, and early operation phase of buildings.

Commissioning and Optimization of Existing Buildings

This area addressed needs for existing buildings and systems to conduct cost-effective commissioning. The focus here was set on existing buildings where the commissioning process must be performed with incomplete or out-of-date documentation.

Commissioning Cost-Benefits and Persistence

This area addressed how the cost-benefit situation can be represented. Key answers were provided by developing international consensus methods for evaluating commissioning cost-benefit and persistence. The methods were implemented in a cost-benefit and persistence database using field data.

Annex 47 was an international joint effort conducted by 50 organizations in 11 countries:

Belgium	• KaHo St-Lieven • Ghent University • Passive House Platform • Université de Liège • Katholieke Universiteit Leuven
Canada	• Natural Resources Canada (CETC-Varennes) • Public Works and Governmental Services Canada • Palais de Congres de Montreal • Hydro Quebec
Czech Republic	• Czech Technical University
Finland	• VTT Technical Research Centre of Finland • Helsinki University of Technology
Germany	• Ebert-Baumann Consulting Engineers • Institute of Building Services and Energy Design • Fraunhofer Institute for Solar Energy Systems
Hong Kong/China	• Hong Kong Polytechnic University
Hungary	• University of Pécs
Japan	• Kyoto University • Kyushu University • Chubu University • Okayama University of Science • NTT Facilities • Osaka Gas Co. • Kansai Electric Power Co. • Kyushu Electric Power Co. • SANKO Air Conditioning Co • Daikin Air-conditioning and Environmental Lab • Tokyo Electric Power Co • Tokyo Gas Co. • Takenaka Corp. • Chubu Electric Power Co. • Tokyo Gas Co., Ltd. • Tonets Corp • Nikken Sekkei Ltd • Hitachi Plant Technologies • Mori Building Co. • Takasago Thermal Engineering Co., Ltd. • Institute for Building Environment and Energy Conservation
Netherlands	• TNO Environment and Geosciences • University of Delft
Norway	• Norwegian University of Science and Technology • SINTEF
USA	• National Institute of Standards and Technology • Texas A&M University • Portland Energy Conservation Inc. • Carnegie Mellon University • Johnson Controls • Siemens • Lawrence Berkeley National Laboratory

FOREWORD

This report summarizes part of the work of IEA-ECBCS Annex 47 **Cost-Effective Commissioning of Existing and Low Energy Buildings**. It is based on the research findings from the participating countries. The publication is an official Annex 47 report.

Report 1, 'Commissioning Overview' can be considered as an introduction to the commissioning process.

Report 2, 'Commissioning Tools for Existing and Low Energy Buildings' provides general information on the use of tools to enhance the commissioning of low energy and existing buildings, summarizes the specifications for tools developed in the Annex and presents building case studies.

Report 3, 'Commissioning Cost Benefit and Persistence' presents a collection of data that would be of use in promoting commissioning of new and existing buildings and defines methods for determining costs, benefits, and persistence of commissioning, The report also highlights national differences in the definition of commissioning.

Report 4, 'Flowcharts and Data Models for Initial Commissioning of Advanced and Low Energy Building Systems' provides a state of the art description of the use of flow charts and data models in the practice and research of initial commissioning of advanced and low energy building systems.

In many countries, commissioning is still an emerging activity and in all countries, advances are needed for greater formalization and standardization. We hope that this report will be useful to promote best practices, to advance its development and to serve as the basis of further research in this growing field.

Natascha Milesi Ferretti and Daniel Choinière

Annex 47 Co-Operating Agents

ACKNOWLEDGEMENT

The material presented in this publication has been collected and developed within an Annex of the IEA implementing agreement on Energy Conservation in Buildings and Community Systems, Annex 47, "**Cost-Effective Commissioning of Existing and Low Energy Buildings**".

This report, together with three companion Annex reports, is the result of an international joint effort conducted in ten countries. All those who have contributed to the project are gratefully acknowledged.

On behalf of all participants, the members of the Executive Committee of IEA Energy Conservation in Building and Community Systems Implementing Agreement as well as the funding bodies are also gratefully acknowledged.

A list of participating countries, institutes, and people can be found at the end of this report.

REPORT EDITORS:
Christian Neumann (Fraunhofer ISE, Germany)
Harunori Yoshida (Kyoto University, Japan)
Daniel Choinière (Natural Resources Canada)
Natascha Milesi Ferretti (NIST, USA)

TABLE OF CONTENTS

1   EXECUTIVE SUMMARY

Building Commissioning (Cx) is a quality assurance process for the design, construction, and operation of buildings. Although recognized as a valuable means to ensure that buildings reach their operating potential, the building commissioning process is not widely adopted internationally. Researchers from Annex 47, an international research project, identified that a major barrier to market penetration is the lack of commissioning methods and tools to ensure that advanced components and systems reach their technical potential and operate energy-efficiently. Results of research to develop guidelines and tools to help overcome that barrier for both existing and future buildings are presented in this report. The results address commissioning and analysis approaches, guidelines for monitoring and the use of sensors, commissioning tools for existing and low energy buildings, and a collection of international case studies.

Commissioning and Analysis Approaches

Although there are international differences in project delivery processes, a general description for building commissioning is useful. Commissioning is generally carried out using either a top-down approach (first evaluating whole system performance and then moving to lower levels), or bottom-up approach (beginning with single components and widening analysis to system level, as needed) and includes the application of relevant performance metrics to evaluate decision points. The needs for commissioning existing buildings are quite different than those for the initial commissioning of advanced and low energy buildings.

The freedom to implement best-practice measures is most easily introduced in the design-phase. However, when commissioning existing buildings, design and operational data is often unavailable, inaccessible, erroneous or outdated. Most building have not been commissioned before and therefore, for existing buildings, commissioning often relies on data that is already available or that can be acquired at low cost. Automating portions of the commissioning process is one approach to improving cost-effectiveness.

In commissioning advanced and low energy buildings, procedures need to explicitly address energy consumption at the system or whole building level and the peak demand that are not typically addressed in current, conventional approaches. Furthermore, new building test procedures for innovative systems are not available in libraries or guides. Advanced and low-energy systems generally require a higher level of control and also greater control accuracy than conventional systems. Innovative buildings are, almost by definition, 'one-of-a-kind' and require functional test procedures that are customized to the design of that unique building.

To improve the building delivery process, research and development must address the need for:

1) Methods to document design intent that extends to integration of systems at the whole building level, and can be clearly understood by the design team, commissioning agents and operators;

2) Design review guidelines for low-energy buildings to help catch problems early;

3) Functional test methods that adequately address innovative system operation and integration issues, and;

4) Functional test methods that compare expected energy performance to actual energy performance during commissioning and diagnose causes of differences.

Commissioning yields the best results if begun early on in the project and integrated into every phase—throughout design, construction, performance verification and acceptance—until stable and optimal operation of the building systems is achieved.

Monitoring Guidelines

Systematic approaches for the evaluation of data, beyond simple benchmarking, are often missing. Furthermore, the unknown benefits versus the costs of installing additional measurement equipment present a major constraint. It is recommended that a project-specific list of necessary measurements and/or sensors be developed by:

- Identifying systems to be assessed;
- Defining specific checks and performance metrics that are appropriate for these systems;
- Deriving the measurements needed to prepare/calculate the desired performance metrics; and
- Defining and using a unified point naming convention, particularly when monitoring massive systems.

In existing buildings, the three most significant problems related to installing sensors were identified as cost, physical constraints from existing facilities, and accuracy. Field experience was distilled to provide proven approaches that help to overcome these problems, including the use of wireless technology and non-invasive sensors.

Commissioning Tools

The diversity of Annex participants provided insight to the needs for commissioning tools from the European, North American, and Asian points of view.

In Europe, energy audits and performance contracts are well established, but third-party commissioning is a new concept, unknown to many consultants and building owners. Although the expertise needed for commissioning is available, when performed, commissioning is generally limited to a handover activity. Standardized descriptions/guidelines and tools that can support a systematic commissioning process are absent.

In North America, barriers to assuring maximum building performance exist at each phase of the building life-cycle: design, construction and turnover, and ongoing operations and occupancy. Cheaper, more accurate and more robust tools to automate the commissioning process for new and existing buildings are needed to improve building performance while reducing labor and other costs. These tools will also help the market better respond to the outsized demand for skilled designers, commissioning providers and operators.

In Japan, the meaning and scope of commissioning varies from engineer to engineer. Pioneering research work has implemented many kinds of commissioning tools, for example, the use of computer simulation to estimate performance of heating, ventilating and air-conditioning (HVAC) systems or sub-systems, and tools to visualize results. Still, adoption is slow. The most significant barriers are the difficulties interfacing with system data, lack of experience and training tools, and lack of robust commissioning tools that can address system interactions.

Across the globe, there is a universal need to measure, monitor, and process huge amounts of data in order to operate the building and systems optimally. Advanced visualization techniques can be used to display the information that is hidden in performance data and/or the recorded operation data of buildings and systems, and, thus, is valuable for commissioning of buildings and systems. Visualization is also relevant for transporting data and information from design analyses, such as simulation and modelling, and to make this information available in subsequent project phases. Visualization techniques can be used to show:

- Change of variables over time;
- Relations between two or more variables;
- Statistical information / distribution of values;
- Spatial information/distribution;
- Partitioning/percentage share of properties;
- Comparison of scalars (elements of different size); and
- Process / information flow.

In Annex 47, a total of eighteen automated and semi-automated commissioning tools were developed. Eight categories were selected to describe the features of existing building and low energy building commissioning tools: objectives, functions, data-management, implementation, operability, analytical engine, end users, and benefits. Detailed features of major tools are listed in Appendix 4. The focus of many of the commissioning tools is fault detection and diagnosis, followed by optimization, design and data handling. The primary target of most tools is system level commissioning, followed by whole building and component level at the same rate, then control level. Most tools have been developed for existing buildings rather than low energy buildings, because automated tools for low energy building (LEB) Cx are still in their infancy.

Case studies were selected from the real building implementation of these tools to present three analysis approaches.

1) Top-down approach - The German ModBen project deals with the systematic performance evaluation of existing non-residential buildings, an ongoing commissioning process. A four-step procedure was developed for ModBen and is illustrated in Figure 1-1.

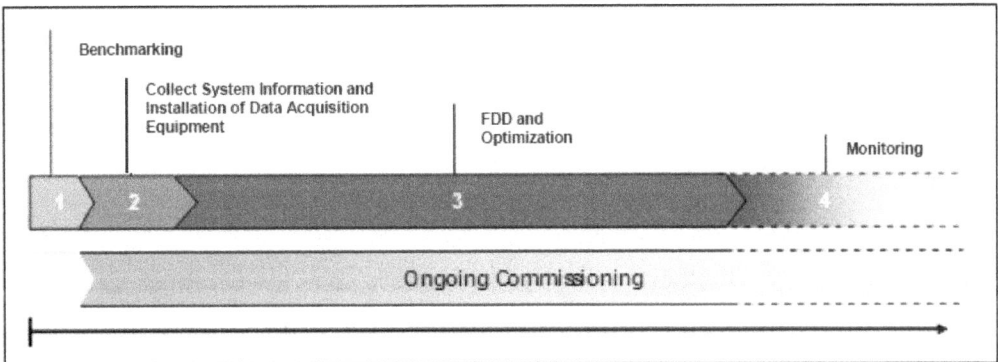

Figure 1-1 The four-step ongoing commissioning procedure on a time scale

2) Bottom-up approach - In the Netherlands, two bottom-up examples were presented. One involved an investigation of tenant complaints and leveraged functional performance testing tools. The second example is a blind analysis of an air-handling unit (AHU) based on a functional description of the AHU as a basis for deriving expert rules, data collected from a building automation system, and no additional sensors.

3) Combined approach - A Japanese promotion program for the introduction of efficient energy systems in housing/buildings assists with the introduction of a building energy management system. In this project, grant-aided businesses specify their demands for achieving energy-conservation. The implementation report includes information on quantitative target values and conditions for their attainment, how to analyze measured data, solve problems under an energy management system consisting of experts, such as a designer and others, in addition to the building's owner, and the procedure for reporting results. This process is considered similar to the commissioning process.

The tools and case studies discussed illustrate the potential for improved building performance through the use of software tools in the commissioning process. The tools provide features for operational fault detection and and/or for optimization that help to identify and realize potential savings.

2 Commissioning and Analysis Approaches for Existing and Low Energy Buildings

2.1 Analysis Approaches

A building and its services can be defined as a hierarchical structure as shown in Figure 2-1. At the highest level, Level 1, a building is seen in its entirety. The lowest level, Level 6, looks at components in a subsystem (e.g., a fan.) The second level is divided into a building's services systems (on left) and the building itself (on right). The systems part consists primarily of physical objects, such as building structure and envelope.

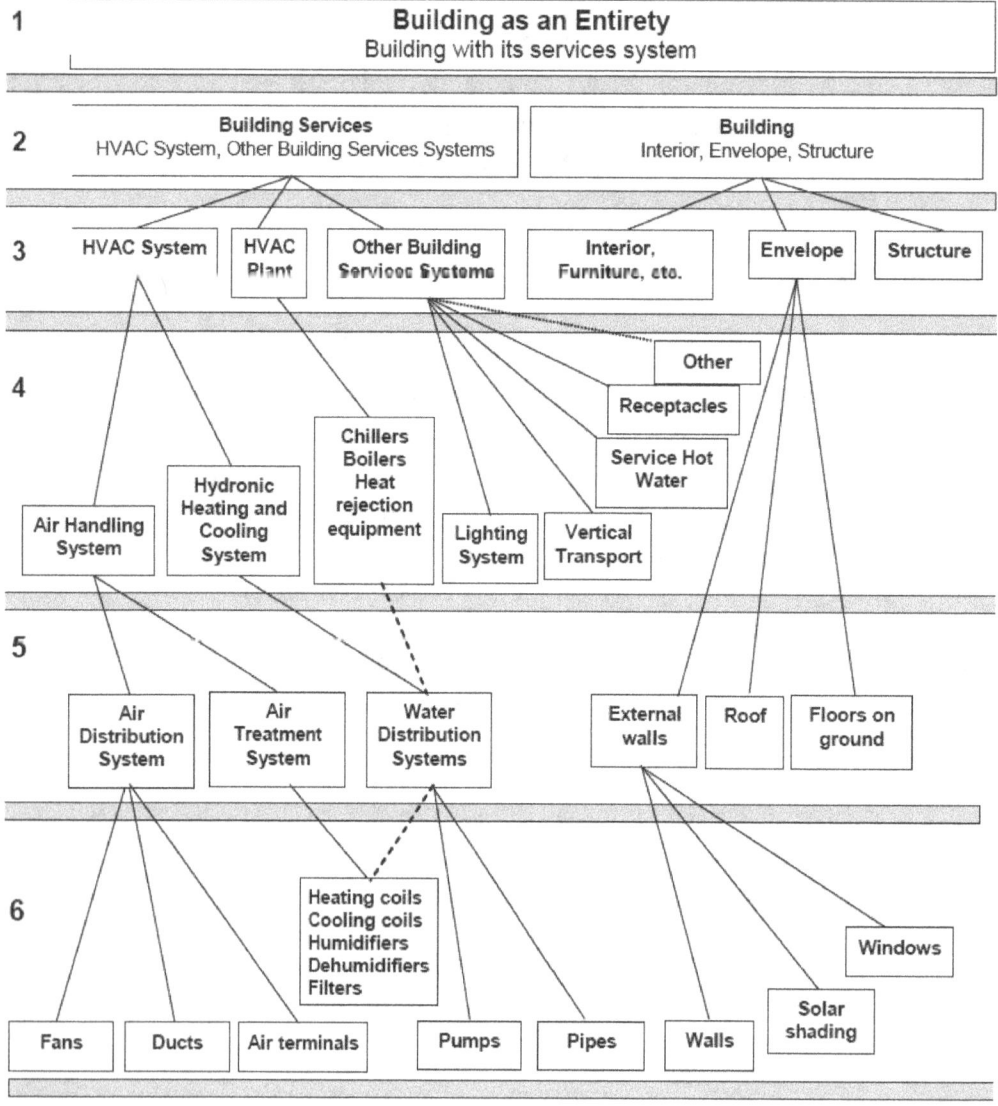

Figure 2-1 Building divided into its service systems, subsystems and components   (Figure source: [1])

4

The overall systems may be viewed from different levels:

- The whole system;
- A subsystem, involving several components; or
- Single components that are considered critical.

As a result, any perspective analysis can occur from a top-down or bottom-up approach. The source book of IEA ECBCS Annex 25 [2] gives these explanations concerning the different approaches.

"The two problems can be presented in the following manner: When undesired operation is observed on the building level, what is the cause of the problem on the level of subsystems or components? When a fault is observed on the component level, what is the seriousness of that fault in terms of building performance on the level of the building as a whole? In Figure 2-2 the problems are represented by arrows. The tail of each arrow indicates an observation of a fault or undesired operation, while the head indicates the result of the reasoning.

"Instead the fault is detected in the building or in the component level; it can as well be detected in the subsystem level. In the last case one must be able to deduce the impact of the fault in the subsystem level in a higher level (building level, for example), and on the other hand, be able to find the cause of the fault in a lower level (component level, for example).

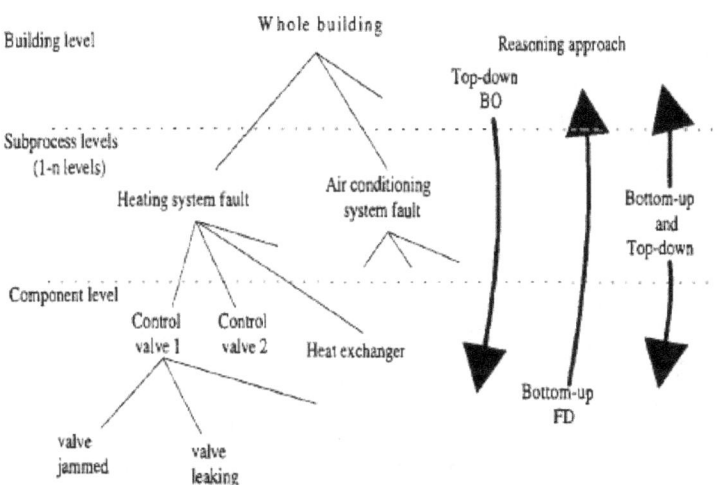

**Figure 2-2   Schema of top-down and bottom-up approaches
(Figure source: [2])**

"Problems can be named according to the principles used for their resolution. In the top down - approach the starting point is some performance property of a building which describes the function of the entire system, such as energy consumption, proceeding towards smaller details. In similar fashion, the bottom up - approach begins with some detail, the significance of which for the whole is not necessarily known, and ending up with the consequence of the fault from the

standpoint of the whole. When starting from the subsystem level, both approaches must be utilized."

2.1.1 Top-Down Approach

The top-down approach consists in "going down" from system level to component level, through some subsystems. The whole system performance is first verified and the inquiry is then extended in turn to lower levels, following observed malfunctions or questions to be answered. The first task is not to verify if a certain component is performing "well" or "poorly", but rather to check if the overall performance of the system and the interaction between components and subsystems is acceptable or within the "normal" range.

The Annex 25 source book [2] also states that:

"What becomes of a problem after the error has been observed, however, is locating it in the building's many subsystems and constituent components. How does localization of the error proceed, and to what degree should it be possible to explain the reason for its occurrence? It should be possible to arrange the reasoning so that it would proceed according to either the order of probabilities or some other corresponding order of importance. For example, if it is a question of how to reduce energy wasting, the first thing that should be considered for inspection is the most probable sub-process which might have been damaged or, alternatively, the partial system which consumes the most energy. If the fault is found there, work should continue according to some preliminary determined system within the subsystem to its components or something similar.

"Localizing the fault in the hierarchical tree would thus proceed to the level possible using existing process data, which has either been obtained from measurements or requested from the user. If there is no control of reasoning of this type, implementation of the reasoning would be too difficult. This leads in practice to a situation in which no resolution of the problem is obtained."

This describes the limitation of the top-down approach: it is well suited to detect (major) faults, but, by definition, not every subsystem and component is monitored. Fault diagnosis could be difficult or even impossible if necessary information is missing. For fault diagnosis, one has to know how a certain fault shows up on high level signals such as the total energy consumption. This could be an easy task for faults such as deficient operating schedules that are not in synchronization with occupancy (demand). However, for more complex faults at subsystem or component levels (e.g., coil fouling, loss of refrigerant or wrongly adjusted operation modes), fault diagnosis could be difficult or impossible and the prognosis is worse when there are multiple faults. Furthermore, the occurrence of a single fault in a component might not evident at high level signals such as energy consumption, but may nevertheless have a significant influence on energy, comfort or lifetime of the component.

The top-down approach is a common solution that combines permanent measurements for the overall performance with temporary measurements at the subsystem or component level for a more detailed analysis of systems that are identified as deficient.

2.1.2 Bottom-Up Approach

The bottom-up approach starts with the analysis of a single component and steps up, progressively, to the whole building level. It might be most appropriate for initial commissioning. It allows a safer identification of local faults, but it might also result in wasted effort because of lack of global vision of the problems.

It is impractical to monitor the operation of all building components. It is therefore essential that, prior to implementation of the bottom-up approach, priorities be established for the investigation of representative samples of those components and faults for which a failure is either probable or would have the greatest effect on the property or set of properties of the building.

2.1.3 Performance Metrics

When assessing building performance it is important that the data are presented in a pre-defined manner that renders it informative. This is achieved using standardized performance metrics.

Deru and Torcellini [3] give the following definitions/characteristics for performance metrics:

"A metric is a standard definition of any measurable quantity, and a performance metric is a standard definition of a measurable quantity that indicates some aspect of performance. Many other terms are used with a similar meaning, such as performance indicator, performance index, and benchmarking.

"Performance metrics need certain characteristics to be valuable and practical. A performance metric should:

- *Be measurable (or able to be determined from other measurements);*
- *Have a clear definition, including boundaries of the measurements;*
- *Indicate progress toward a performance goal; [and]*
- *Answer specific questions about the performance."*

This definition is applied to energy performance of buildings and performance metrics for different levels according to Figure 2-3. Deru and Torcellini defined three levels of performance metrics: Tier 1Metrics, Tier 2 Metrics, and Indicators.

Tier 1 metrics provide a high level building overview e.g., from monthly/annual data such as utility bills. Tier 2 metrics provide a breakdown of the performance of subsystems and/or components and typically requires hourly or sub hourly data. Indicators are above Tier 1 metrics, and they aggregate complex information to show planning level trends toward goals.

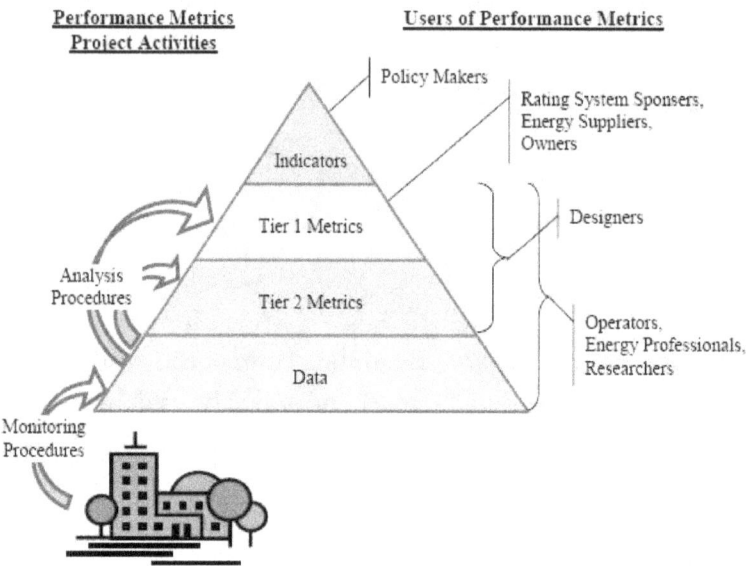

Figure 2-3 Application of performance metrics in building performance evaluation[3]

Even though Deru and Torcellini focused on energy performance metrics (based on measurements of energy flows) the definition of performance metrics and tiers can be applied to any general analysis that incorporates any kind of measurable data (see also: Gillespie, et al.[4]). In general, the definition of performance metrics follows the identification of systems to be monitored and questions to be answered concerning their performance

2.2   Special Aspects of Existing Buildings

For new construction, ongoing commissioning may be introduced in the design phase, giving the building owner and design engineer freedom in their decisions. With existing buildings that were never the subject of any kind of commissioning, this freedom is missing. The design and structure of the building and the HVAC system are fixed. Introducing ongoing commissioning in existing buildings poses two significant challenges: 1) lack of design data and 2) lack of sufficient budget to collect robust data.

For existing buildings, information about the construction and systems may be unavailable, de-centralized, inaccessible, or contain erroneous or invalid documentation.   Metering data is normally reduced to the minimum necessary for the energy billing.

Although it is feasible to prepare detailed documentation, create building modeling and/or install numerous sensors to measure data for the analysis, the cost is deemed too high for this to be standard procedure in existing buildings. The budget is a significant factor when considering measurement equipment as well as the effort required for acquisition of information during the audit.

As a result, tools for existing buildings often rely on data which is either already available or that can be acquired at low cost. Consequently, the choice of approach used in analysis of an existing building (top-down or bottom up) often depends on the availability of measured data. The analyst may combine the following complementary approaches:

1. Use existing sensors only
    This approach tries to make "the best" of the situation. An example of a top-down-approach would be benchmarking based on utility bills (monthly/annually).

2. For a bottom-up analysis

    This approach is only reasonable if a building automation system (BAS) is installed that can provide a lot of data from the system. As the data from the BAS might not be sufficient to analyze every component – the outcome of this approach might be arbitrary or incomplete.

3. Use and/or install a consciously chosen set of sensors.
    This approach tries to use or install only necessary sensors to answer specific questions while adhering to budget requirements.

In practice, a mixture of these approaches will likely be applied as listed in Table 2-1.

Table 2-1 Combination of top-down/bottom up approaches and use of existing/new sensors

	top-down	bottom-up
Use existing sensors only	Benchmarking based on utility bills/meter readings (monthly / annual)	Analysis based an data provided by existing BAS
Use and/or install a chosen set of sensors	Detailed monitoring of building's energy balance.	Detailed monitoring and fault detection and diagnosis (FDD) of building systems.

While a top-down approach (that is more than simple benchmarking) often starts with an energy balance and likely requires new sensors, a bottom-up approach often works with existing sensors to avoid high costs for measurement equipment.

Furthermore, a top-down approach is better suited to continuous, long-term evaluation of overall building performance, while a bottom-up approach should only be used to track problems of single components with a temporary measurement.

In general, both approaches are necessary to achieve optimum performance because the top-down approach might not detect or diagnose problems at the component level, while the bottom-up approach cannot make an assessment of the overall energy consumption.

The deployment of sensors in buildings is discussed in more detail in Chapter 3.

2.3 Initial Cx of Advanced & Low Energy Building Systems

2.3.1 General Characteristic of Advanced & Low Energy Systems

As buildings and systems become more and more integrated and advanced to strive for better performance and higher energy efficiency, more attention must be given to the dynamic operation and system interactions. Modern buildings use advanced technology to reduce the energy demand. Systems are designed and sized close to their optimal operation point to provide the highest efficiency. Furthermore, systems are integrated to avoid contradictory or even conflicting operation modes, as well as to leverage synergies. Examples include advanced room climate concepts with integrated controls for lighting, daylighting, heating and cooling; energy recovery in air and water systems; co- and tri-generation.

Another trend in advanced buildings is to use the building façade and thermal building mass to provide better comfort in occupied zones. Strategies include passive as well as active elements, such as natural ventilation; night time ventilation for pre-cooling a building; use of solar gains for heating; etc. Radiant heating and cooling systems have the potential to provide a high level of comfort with a low energy demand but require particular attention when it comes to the proper operation and control. The same is true for low energy systems in general, since these systems are operated at temperatures close to the demand to provide better efficiency on the generation side and avoid energy losses on the distribution and terminal side.

Generally, advanced and low-energy systems require a higher level of control and also greater control accuracy than conventional systems. Not only do they operate with a higher dynamic, but also within a relatively narrow bandwidth where they perform most efficiently. Due to intensive interaction between systems, required operations and any interference with it must be based on sound knowledge of the consequences of any action.

Commissioning for advanced and low-energy buildings must therefore be a planned, systematic quality control process with the objective to ensure that all the building's energy related systems are installed and calibrated, and perform interactively according to the design intent and operational needs as defined in the owner's project requirements (OPR), and specified in the basis of design (BOD), and construction documents (CD).

The scope of the commissioning process follows, in detail, the entire project delivery process. At the beginning, the OPRs give a general outline of the project expectation, usually on a building level. Typical requirements include function and functionality, energy efficiency, sustainability goals, and, not least, cost. With the BOD, the design team develops solutions on a system level, such as recommendations and design criteria for HVAC systems, the building envelope, and the basic concept for controls and building automation. The level of detail then reaches the component level during design development and the preparation of specifications. The commissioning process follows this gradual development with reviews of the associated design documentation.

Then, as construction commences, checks and verification begin at the component level, continue through the system level, and finally are undertaken at the general building level once again. This strategy ensures efficiency in execution, completeness of the review and verification process, and that associated prerequisites are met for each testing phase. Figure 2-4 illustrates the general commissioning process and the focus and level of detail for each step.

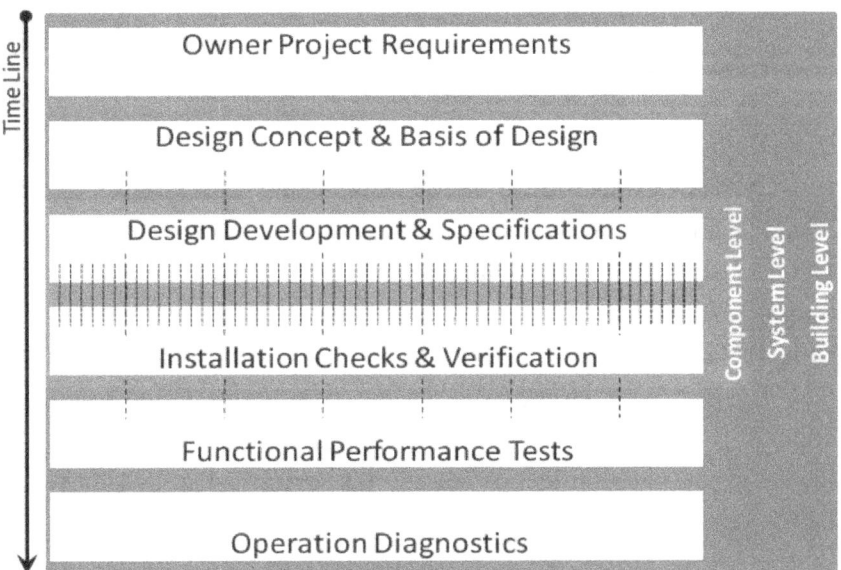

Figure 2-4 An overview of the general commissioning process

It is critical that owners, design professionals and contractors be involved in this quality control process, as well as users, occupants, and operations and maintenance staff. Since not all project team members are available throughout the entire project duration, it is the Commissioning Agent's (CxA) function and main responsibility to facilitate and establish communication, understanding, and information flow between them. Commissioning yields the best results if begun early on in the project and integrated in the delivery process throughout design, construction, performance verification and acceptance until a stable operation of the building systems is achieved.

Commissioning during the design phase is intended to achieve the following specific objectives:

- Ensure that the design and operational intent are clearly documented;
- Ensure that recommendations for improvements are communicated to the design team during design to aid the development of commissioning and avoid later contract modifications; and
- Ensure that commissioning for the construction phase is adequately reflected in the construction documents.

Commissioning during project construction is intended to achieve these specific objectives:

- Ensure that applicable equipment and systems are installed properly and receive adequate operational tests by installing contractors;

- Verify and document proper performance of equipment and systems through normal and other likely operational modes necessary to meet design intent;
- Ensure that O&M documentation provided to Owner is complete; and
- Ensure that the Owner's operating personnel are adequately trained.

2.4   References Chapter 2

[1] Nilsson, P -E. (Ed), 2003. Achieving the Desired Indoor Climate – Energy Efficiency Aspects of System Design, Studentlitteratur, the Commtech Group, Lund, Sweden.

[2] J. Hyvarinen, Karki S (eds.) « Final Report Vol 1: Building Optimization and Fault Diagnosis Source Book » Finland, Technical Research Centre of Finland, Building Technology, 1996, 324 pp.

[3] M. Deru and P. Torcellini, Performance Metrics Research Project – Final Report", Technical Report, National Renewable Energy Laboratory, USA, NREL/TP-550-38700, October 2005.

[4] K. Gillespie, et.al., "A Specifications Guide for Performance Monitoring Systems", CEC PIER-DOE project, USA, 2007.

3 Monitoring Guidelines and Deployment of Sensors for Existing Buildings

This chapter gives a general overview of problems connected to the monitoring and deployment of sensors in existing buildings and some typical problems related to their installation and accuracy.

3.1 Introduction

The following statement is taken from Kenneth Gillespie et al. [5].

"Those who evaluate the performance of buildings and their energy using systems have long known that it takes the attention of a knowledgeable and dedicated team to obtain the quality of data necessary to determine how well a building is actually performing as well as identify means for improving it. This team may include a measurement analyst, instrumentation vendors, an installation contractor and the owner's staff. The problem is that buildings are not designed for measuring their performance. This is particularly true of flow. It is also believed that obtaining such data is a luxury; that it is not needed for system control or day to day operations."

This is not only true for existing buildings in the U.S. but likely describes the situation in many countries around the world.

It is important to understand that the problem described above comprises more than just missing sensors and measurements. Systematic approaches for the evaluation of data that go beyond simple benchmarking are often missing. Therefore, it is often unclear to the building owner or operation staff which measurements are necessary. Furthermore, the unknown benefits and costs of installing additional measurement equipment present a major constraint.

Ideally, the steps to develop a list of necessary measurements or sensors can be described as follows:

1. Identify system to be assessed.
   This could be the whole building, a subsystem (e.g., air handling unit), or a single component (e.g., heat pump). At the same time, the system limits (in the sense of control volume) have to be defined.
2. Define specific questions related to these systems.
   E.g., "What is the system's COP?", "Is the building's specific daily energy consumption in an acceptable range?", or "Does the operation mode of the air handling unit fit the actual limit conditions?"
3. Define performance metrics that can validate the performance of the system.
   This could be anything from simple calculations to complex evaluation routines, (e.g., from the definition of a generator's COP to the automated outlier detection for a building's heat consumption).
4. Derive necessary measurements to prepare/calculate performance metrics.

5. Adopt a unified point naming convention. This is especially helpful for massive systems.

Thus, the need for measurements depends on the system being considered and the questions to be answered, and, in turn, this choice of systems and questions is often highly dependent on the available budget. For example: A simple benchmarking for a building's total yearly energy demand requires fewer measurements and less budget than comprehensive performance monitoring of a complex air handling unit. Temporary measurements can cost less than permanent installations while answering the same questions.

Compared to new construction, existing buildings pose additional constraints. For new construction, monitoring may already be considered during the design phase and thus be integrated in planning HVAC systems and building energy management system (BEMS). In existing buildings this is not the case and monitoring often must build on the given system which is generally not geared for performance monitoring. Depending on the available measurement sensors and data handling equipment, the requirements and cost for additional installations can vary widely. Furthermore, in existing buildings, problems typically arise from the HVAC system's structure that can make additional installations difficult and even impossible.

Hence, for existing buildings, the commissioning provider and building owner and/or operations staff must agree on the level of detail for the analysis (i.e., choice of systems and questions) that correspond to a given budget. From there, a sensor list can be developed. Some iterations might be necessary to adjust the monitoring system to the budget.

This task is specific to the building and/or building owner. However, as buildings and HVAC systems consist largely of similar systems, subsystems and components, the question may arise whether it is worthwhile to provide general guidelines for performance monitoring of existing buildings, at least for the most typical systems.

3.2 Need for Monitoring Guidelines and Present Status

A clear experience for many investigators shows that many buildings have only main energy meters if at all. Typical questions to ask when assessing building performance are: How much energy is used and for what? What are the operating conditions (weather, schedules, set points for temperatures and flow rates)?

A lot of individual approaches for the deployment of sensors have been developed in practice and in the scientific community (also see Chapter 2). However, more general and systematic compilations of necessary measurement points and evaluation routines for the assessment of the performance of typical systems (i.e., monitoring guidelines) are rare.

Ideally, general monitoring guidelines should include the following for a building's typical systems, subsystems and components or HVAC system respectively:

- List of necessary measurements;
- Definition of the kind and quality of data acquisition;
- Definition of type and quality of data storage and data handling;

- Definition of necessary time resolution and duration (permanent or temporary) of measurements;
- Definition of performance metrics and typical benchmark values;
- Definition of a unified point naming convention; and
- Examples of applications of the guidelines.

This list should be defined further for various levels of detail concerning the analysis to account for varying budgets.

Today there are only few examples for general monitoring guidelines that cover points mentioned. Examples of such guidelines are provided in 3.2.1-3.2.5.

3.2.1 SHASE Energy Performance Measurement Manual for Building Equipment & System–Japan

SHASE (The Society of Heating, Air-Conditioning and Sanitary Engineers of Japan) published the "Energy Performance Measurement Manual for Building Equipment & Systems."[6] This manual (written in Japanese) organizes views on performance measurement methods for air-conditioning facilities and plumbing sanitary systems, and defines standards of recommended measurement points for evaluation, deployment of sensors, code name of measurement points, evaluation index, and tips to evaluate the performance for each sub-system.

3.2.2 "A Specifications Guide for Performance Monitoring Systems", University of California Lawrence Berkeley National Laboratory – USA

The Specifications Guide for Performance Monitoring Systems [7] defines three different levels of performance monitoring (basic, intermediate and advanced) with increasing levels of detail of the analysis but nonetheless comprising the whole building or HVAC system. The performance metrics, the measurement system requirements, the data acquisition and archiving and the data visualization and reporting are defined for each level. The type and accuracy of sensors are provided as well as a unified point-naming convention.

3.2.3 "EnOB Monitoring Guidelines" – Germany

The EnOB Monitoring Guidelines (written in German) specify the requirements for performance monitoring in the German EnOB program [8], which is a loan program for the design and construction of low energy buildings. It defines measurement system requirements and some overall performance metrics. It gives only one level of detail, that being on the data acquisition system but provides no information on accuracy.

3.2.4 IPMVP

The International Performance Measurement and Verification Protocol (IPMVP) [9] describes concepts and options for determining energy and water savings in buildings. Development of the protocol is sponsored by the U.S. Department of Energy (DOE) and an international coalition of facility owners/operators, financiers, contractors and Energy Services Companies (ESCOs). It gives four options for calculating energy savings:

A: Partially measured retrofit isolation

Savings are determined by partial field measurement of the energy use of the system(s) to which an energy conservation measure (ECM) was applied; separate from the energy use of the rest of the facility. Measurements may be either short-term or continuous. Partial measurement means that some but not all parameter(s) may be stipulated.

B: Retrofit isolation

Savings are determined by field measurement of the energy use of the systems to which the ECM was applied; separate from the energy use of the rest of the facility. Short-term or continuous measurements are taken throughout the post-retrofit period.

C: Whole Building

Savings are determined by measuring energy use at the whole facility level. Short-term or continuous measurements are taken throughout the post-retrofit period.

D: Calibrated simulation

Savings are determined through simulation of the energy use of components or the whole facility. Simulation routines must be demonstrated to adequately model actual energy performance measured in the facility. This option usually requires considerable skill in calibrated simulation.

Depending on the availability of historical consumption data and the kind of energy saving measure that is to be implemented, various approaches may be chosen. Option A and B deal with single measures that can be separated by (partial) measurements and thereby determine the savings that can be gauged. Option C and D refer to the whole building level, so multiple measures can be evaluated using these options. However, there is no further specification of measurement or performance metrics.

3.2.5 CEN Standards - Europe

Some European standards encourage use of measurements for their implementation. These standards are developed for performance estimation of equipment. The standards can be categorized as follows:

Requirements for Inspection

Standards EN 15239:2007 [10], EN 15240:2007 [11] and EN 15378:2007 [12] provide instruction for inspection of ventilation, air-conditioning, and boiler and heating systems, respectively. These standards support essential requirements of EU Directive 2002/91/EC on the energy performance of buildings (EPBD) [13]. In EN 15240:2007 it is noted that energy consumption or running time meters may have been installed on air-conditioning systems. Regular notation of the meter readings can help assess operation of the air conditioning system. Standard EN15378 lists which measurements are required or optional depending on plant capacity. All three standards list what should be checked during inspection, however, measurement equipment and measurement accuracy have not been defined. Therefore, a guideline on possible measurement equipment for inspection purposes could be useful.

Advice following inspection

After completion of plant or equipment inspection, recommendations for possible improvement must be included in the inspection report. Some of these recommendations imply use of new measurements. For example, the standard for the inspection of air-conditioning systems, EN 15240:2007 [10], states that if there is no energy metering in place, a recommendation would be to install appropriate energy consumption metering to the largest energy consuming air-conditioning plant, and subsequently to record the consumption on a regular basis. In addition, the standard for inspection of boilers and heating systems [11] lists possible improvements that include measurement of energy use. Although an inspection report includes recommendations regarding measurement, there is no advice on sensor type, measurement accuracy and integration.

Testing

Standards EN 308:1997 [14], EN 12599:2000 [15], EN 13829:2000 [16], and EN 13187:1998 [17] relate to the testing of components. For example, standard EN 12599:2000 [15] is for testing ventilation and air-conditioning systems. The standard explains measurement accuracy and different methods together with equipment location for measuring air flow .Standard EN 13829:2000 [16] gives instruction on how to measure the amount of air infiltration using pressure differential. In this standard measurement, accuracy and equipment location are also given. To measure thermal irregularities in building envelopes, a sensor for infrared radiation must be used, as mentioned in Standard EN 13187:1998 [17]. In this standard, measurement accuracy for theinfrared method is given. As these four standards show, standards for testing and measurement require detailed instruction and requirements for accuracy of measurement. Therefore a new guideline for sensors and measurements must include sensor description, so that a proper sensor can be chosen for a given type of measurement.

Requirements for Measurement

In the standard for performance requirements for ventilation and room-conditioning systems, EN13799 [18], it is mentioned that the design criteria constitutes the basis for measurements that will be carried out during the hand-over process. In addition, this standard gives general requirements for control and monitoring, and states that a measuring concept shall be identified at an early stage of the project and necessary measuring devices installed. Even though this standard gives general requirements for measurement definition in the design phase, there is no specification on measurement equipment and accuracy. Therefore a guideline for sensors can be a useful reference to this standard on performance requirements, to support sensor deployment in the design phase.

Common Metrics

All nine standards mentioned can be used for performance verification, inspection, or measurement. Regardless of the standard purposes, there are some common metrics to be measured. For example, in standards EN 308:1997 [14], EN 15239:2007 [10], and EN 12599:2000 [15] air flow rate should be measured. In some standards there are parameters that should be measured indirectly by measuring other parameters, like equipment efficiency or Specific Fan Power (*SFP*, a parameter that quantifies the energy-efficiency of fan air movement systems) for example. Since there are common metrics that should be measured directly or indirectly in different standards, a guideline on sensor deployment as a reference to the above standards can encourage use of a particular sensor for a given application.

3.2.6 Recommendations

From the experience of the Annex 47 team it seems worthwhile to further develop such systematic approaches and publicize them for general use. For certification systems such as Leadership in Energy and Environmental Design (LEED) (USA), *Building Research Establishment Environmental Assessment Method (BREEAM) (UK), Deutsche Gesellschaft für Nachhaltiges Bauen* (DGNB) (Germany) or CASBEE (Japan), it would be interesting to integrate systematic monitoring guidelines.

3.3 Requirements and Problems for Sensor Deployment in Existing Buildings

3.3.1 Problems Installing Sensors in Existing Buildings

Sensing systems are essential for measuring the performance of constructed facilities. Nevertheless, awareness of performance evaluation is low, and construction cost constraints often limit the number of permanently installed sensors. This has led to there being too few sensors in existing buildings.

To grasp the performance of such existing buildings requires the installation of additional sensors. The installation of sensing systems in existing buildings involves the following problems:

- Limited budgeting for the installation of a sensor system;

- Insufficient space and other constraints in existing facilities; and

- The accuracy limitations of added sensors. Kao and Pierce [19] found that sensor errors can increase the annual energy requireents attributable to an air handling system by as much as 30 % to 50 %.

3.3.2 Results of Survey on Sensor Deployment in Existing Buildings

To understand the current situation, we conducted a survey on sensors installed in existing buildings to investigate the four points listed.

1) Is the number of sensors installed adequate for energy management?

2) Can additional sensors be installed notwithstanding restrictions such as installation space?

3) Are the sensors sufficiently accurate for energy management?

4) Does installation of additional sensors for energy management require any action by the building owners, such as budgeting measures?

The number of responding design offices, general contractors, construction companies, consulting companies and others surveyed are:

Asia: Japan 60

North America: United States 6, Canada 1

Europe: Germany 1, The Netherlands 1, Belgium 1, Norway 3

The response breakdown is shown in Table 3-1, while the results summary is presented in detail in Appendix 1.

Table 3-1 Response rates

	Asia (Japan)	North America	Europe
Design offices	15 %	0 %	0 %
General contractors	23 %	17 %	0 %
Construction companies	38 %	0 %	20 %
Consultants	5 %	33 %	60 %
Others	20 %	50 %	20 %

1) Sensor installation situation

The following categorization of Building Energy Management Levels (A to F) was proposed by the SHASE BEMS Technical Committee and it was used in our investigation of the current state of energy level management.

A. The total amount of energy can be computed for the entire building. (type of energy; monthly units)

B. The total amount of energy can be computed for the entire building. (type of energy; daily and hourly units)

C. Energy can be measured or computed by type of use. (e.g., lighting, air conditioning, satellites, elevators, etc.)

D. Energy can be measured or computed by system. (e.g., office system, conference room system, executive office system, computer room system, parking area system, etc.)

E. Energy can be measured or computed by floor.

F. Energy can be measured or computed for specific machines or subsystems. (e.g., cooler, heating system, etc.)

Results show that it is currently possible to compute the total amount of energy for the entire building for almost all buildings (level A). Nevertheless, less than 50 % of buildings have an energy management level B, and levels C through F account for less than 30 %. Looking at the proportions for levels C to F by region, we see that Japan has a higher proportion of high energy management level buildings, (15 % to 30 %) than Europe (5 % to 15 %) or North America (10 % or less).

Furthermore, the survey's respondents thought measurement by use (level C) or daily or hourly measurement for the entire building (level B) should be considered, wherever possible. Most respondents in favor of level C considered investigation of energy conservation to be impossible without measurement by use. Level B respondents considered short-period (hourly) minimum sensor measurements to be optimum for obtaining energy consumption trends and profiles, taking cost-benefit into consideration.

Investigation of problems with the sensing system also produced responses of 'not enough sensors' from nearly all respondents, leading to the conclusion that few buildings have a sufficient number of sensors installed.

2) Sensor installation space

Existing buildings in which there is sufficient space for installation of sensors is 52 % for North America, 39 % for Europe, and 16 % for Japan. The problem of space for sensor installation is particularly prevalent in Japan, because machine rooms are designed to be extremely compact.

3) Sensor accuracy

Very few responses regarding sensor accuracy were obtained. We believe that either the respondents had low concern for sensor accuracy or had insufficient knowledge of sensor accuracy to respond.

4) Sensor types and budgeting for additional sensor installation

The results of investigating budgets for the installation of additional sensors for the energy management levels of items revealed that most budgets approved attainment of level A but full approval was 50 % or less for level B, and full budgeting for levels C through F was 30 % or less.

The proportion of full approval for levels C to F is higher in Japan (18 % to 30 %) than in North America (20 % or less) or Europe (20 % or less). The survey found building owners in Japan were more concerned with energy management than those in Europe and North America.

3.3.3 Requirements for Installing Sensing Systems in Existing Buildings

From the survey results, we conclude that although existing buildings have inadequate numbers of sensors, budgets and space are problems for the installation of additional sensors. Problems for additional sensor installation and countermeasures are listed in Table 3-2.

Table 3-2 Problems and countermeasures for installing sensors in existing buildings

Problem	Countermeasures		
	Sensors	Transmission System	Data Processing
Cost	- Fewer sensors - Simple and quick Installation - Develop simple calibration tools	- Wireless or power line transmission	- Use of automated software tools to check for wiring errors - Use of sensor data built into devices
Constraints in Existing Facilities	- Select ultrasonic flow meters (no need to cut ducts) - Select pressure differential flow meters (work with short-run ducting)	- Wireless transmission	- Open processing via gateway (BACnet, etc.) (communication protocol)
Accuracy	- Select integrating calorimeters (calibrated with difference in readings of two temperature sensors)		- Correction of measured data with relative values

1) Cost

To conduct a detailed performance evaluation, the number of sensors is increased to match the desired degree of detail. That, however, is accompanied by an increase in cost. The additional cost is not only for hardware; high labor costs for installation are included. The following are ways to reduce the total cost of the sensing system.

- Develop tools for checking wiring for errors
- Develop simple calibration tools
- Use data from sensors built into machinery
- Use wireless communication or power lines to transmit data (reduce wiring cost)
- Design for future additional installations
  - Centralize installation at location of power meters
  - Install ducting with straight-run lengths for installation of flow meters
- Reduce number of sensors (Use minimum number of sensors for the purpose; standardize sensor placement.)
- Simple and quick sensor installation

21

2) Constraints in existing facilities

When installing sensors in existing buildings, certain constraints must be overcome. Examples of new inventions to overcome these constraints are:

- Sensors
    - Flow meters such as ultrasonic flow meter that can be installed without cutting ductwork
    - Flow meters that can be installed within a small space (a short straight-run length) such as a special sensor that computes flow rate from the relationship between flow speed and pressure drop

- Transmission system
    - Dispersion of sensor locations to reduce wiring cost by using wireless communication, power line wiring communication, etc.

- Communication protocol
    - Data communication with existing BAS, machines, etc. by implementing open protocol (BACnet, LON, etc.)

- Accuracy

Under cost, space and other constraints, it is necessary to ensure that accuracy is adapted to purpose. The way in which overall accuracy of the sensing system is attained is illustrated in Figure 3-1.

Sensing	Transferring	Analog Conversion	A to D Conversion	Digital Data Treatment
Sensing Error/ Output Error	Cable Resistance	Amplifying Error	Linearity Error	Digit Trouble
V fluctuation	-	V fluctuation	Sampling Error	Rounding Error
Noise	Noise	-	Holding Error	-
Temperature Drift	-	Temperature Drift	-	-

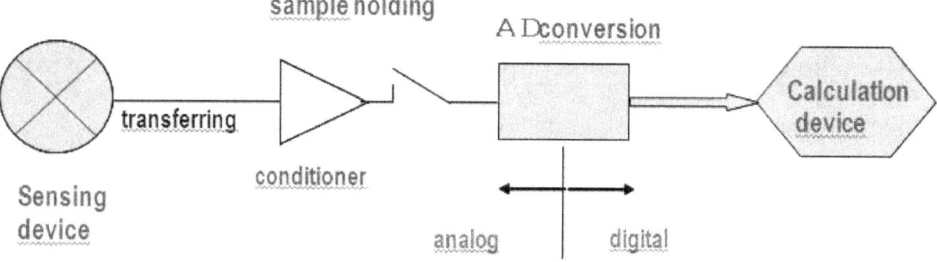

Figure 3-1 Effect of hidden errors on overall accuracy in the monitoring system

Errors in the sensing system stem from analog errors in the sensing elements and wiring, and conversion errors when converting the digital signals to analog format. In addition, the measured values from two temperature sensors in a flow meter are used in computing the amount of heat, multiplying individual errors. Thus, when performing calibration to correct errors, it is important to calibrate the sensing system as a whole. (When the combinations of sensor elements, wiring, AD converters, etc. are changed, the overall characteristics also change, so re-calibration of the new combinations becomes necessary.) A detail example of error in sensing system is presented in Appendix 2.

Compatibility with existing sensors is also important. Values relative to existing sensors and sensor validity and correction, etc. should be checked. When measuring temperature difference with sensors installed at the intake and outlet of a refrigeration unit for example, it is important to check sensor accuracy with the refrigeration unit turned off using the difference in temperature readings of the two temperature sensors to correct the output value.

3.4   References Chapter 3

[5] Kenneth, Gillespie, et. al., "A Specifications Guide for Performance Monitoring Systems", CEC PIER-DOE project, USA, 2007.

[6] Society of Heating, Air-Conditioning and Sanitary Engineers of Japan) published the "Energy Performance Measurement Manual for Building Equipment & Systems." http://www.shasej.org/English/

[7] University of California, Lawrence Berkeley National Laboratory "A Specifications Guide for Performance Monitoring Systems" Specifications Guide Version 1 3/23/2007

[8] EnOB Monitoring Guidelines, http://www.enob.info/en/research-areas/enob-key-research-areas/monitoring-and-analysis-accompanying-research-for-model-projects/ (website updated November 2010).

[9] IPMVP Technical Committee. 2010. International Performance Measurement &Verification Protocol Volume 1: Concepts and Options for Determining Energy and Water Savings. U.S. Dept. of Energy: 86.

[10] EN15239, Ventilation for buildings, Energy performance of buildings, Guidelines for inspection of ventilation systems, Bruxelles, 2007.

[11] EN15240, Ventilation for buildings, Energy performance of buildings, Guidelines for inspection of air-conditioning systems, Bruxelles, 2007.

[12] EN15378, Heating system in buildings, Inspection of boilers and heating systems, Bruxelles, 2007.

[13] The Energy Performance of Buildings, Directive 2002/91/EC of the European Parliament and of the Council, European Union, 2002.

[14] EN308, Heat exchangers, Test procedures for establishing performance of air to air and flue gases heat recovery devices, Bruxelles, 1997.

[15] EN12599, Ventilation for buildings, Test procedures and measuring methods for handling over installed ventilation and air-conditioning systems, Bruxelles, 2000.

[16] EN13829, Thermal performance of buildings, Determination of air permeability of buildings, Fan pressurization, Bruxelles, 2000.

[17] EN13187, Qualitative detection of thermal irregularities in building envelope, Bruxelles, 1998.

[18] EN13799, Ventilation for non-residential buildings, Performance requirements for ventilation and room-conditioning systems, Bruxelles, 2007.

[19] J.Y. Kao, E.T. Pierce, Sensor errors: their effects on building energy consumption ASHRAE J. (1983), pp. 42–45 Dec.

4 Cx Tools for Existing Buildings and Low Energy Buildings

4.1 Needs for Cx Tools

4.1.1 Needs from the European Point of View

The term "commissioning" is less prevalent in Europe – though many services offered do relate to commissioning. Furthermore, the situation in most countries is very different for existing buildings than for new construction.

For existing buildings, most companies and consultants offer third-party services ranging from energy concepts to energy audit and performance contracting. However, no general systematic approach or standardization exists yet. These services are usually individual approaches based on a company's expertise and knowledge; so consequently, no tools are distributed or used (except common spreadsheet programs).

For new buildings the situation is different as commissioning is typically not offered as a third-party service in most countries. Furthermore, in most cases commissioning is only performed during the hand-over as initial commissioning. This is usually performed by the builder or planner. In many cases this initial commissioning is only a rough check of completeness of installation and principal functionality. A commissioning process as described in IEA ECBCS Annex 40 [20] for new constructions is not at all usual. Tools for a systematic and more standardized approach are also absent.

Discussion of energy efficiency in buildings in Europe grew significantly after the European Performance of Building Directive (EPBD) came into force [21]. (The Council of the European Union passed directive 2002/91/EC in December 2002.)

The purpose of the directive is to make energy use in buildings more effective and by that contribute to reducing greenhouse gases. This can significantly reduce dependence on energy and provide improved security.

The directive determines requirements for:

- A calculation methodology for building energy performance;
- Minimum requirements regarding energy performance of new buildings and buildings comprehensively refurbished;
- Energy performance certification;
- Regular inspections or advisories regarding heat boilers; and
- Regular inspections of HVAC systems.

In light of this list, the EPBD agrees in principle with the concept and supports commissioning. Unfortunately, current member states have not implemented this for potential use.

While commissioning extends over the entire life of a building, certification according to the EPBD will likely only be performed once. The EPBD does not prescribe an ongoing evaluation of building performance. Accordingly, certification can only be one part of the ongoing commissioning process.

In the case of asset ratings (theoretical calculation of the energy demand), certification can deliver the actual building status and a theoretical target value for energy performance. Thus, it could be integrated in phase I of the development of the ongoing commissioning process.

However, asset ratings for existing buildings are currently only prescribed in Denmark and Austria. Most Member States will have operational ratings (based on utility bills) for existing buildings, which as currently defined, are not suited for any detailed analysis.

The only systems explicitly dealt with in the EPBD are the inspection of air-conditioning systems and boilers. The last is implemented nationally, either as inspections or as information, depending on each country's need. Again, the kind of inspection and the intervals in most Member States are not suited to the ongoing commissioning process.

For more information, see (www.buildingsplatform.org or www.buildingeq.eu)

Generally speaking, commissioning in Europe is currently a service offered by a few experts only; many are researchers in that field.

However, a lot of the expertise needed for commissioning is available. But neither consultants nor customers are familiar with the concept of commissioning. Standardized descriptions/guidelines and tools that can support a systematic commissioning process are absent.

4.1.2   *Needs from the North American Point of View*

Cheaper, more accurate and more robust tools to automate the commissioning process for new and existing buildings will improve building performance while reducing labor and other costs. These tools will also help the market better respond to the need for skilled designers, commissioning providers and operators.

In North America, barriers to assuring maximum building performance exist at each phase of the building life-cycle:

- Design: Owners and designers often ignore the benefits of design stage commissioning activities, including documentation of design intent and design review to verify constructability, testability, operability and maintainability. There is a lack of tools for design phase commissioning.
- Construction and turnover: Owners often perceive commissioning during construction as an unnecessary cost. There are insufficient tools to streamline the labor-intensive processes of testing and integrating multiple building systems to reduce costs. There is no accepted procedure for short term testing to compare energy performance against requirements and expectations developed during the design phase.

- Ongoing operations and occupancy: Few maintenance staffs have the time, tools, or training to perform the ongoing assessments of building performance required to assure the persistence of energy savings obtained during construction phase commissioning or retro-commissioning. Simple or automated diagnostic tools are necessary to make ongoing assessment easier and more cost-effective and reduce or eliminate the need for repeated retro-commissioning.

Four areas of tool development needs were identified by a commissioning and diagnostics research team funded by the U.S. Department of Energy [22].

1. Design Intent Documentation Tool

The documentation of design intent is a key element of commissioning, and one often overlooked. It is an effective means of communicating important information to operating staff, and others, who might not otherwise receive it. Design intent documentation also provides a useful communication tool among members of the design team and between the team and the building owner.

There is a need to provide tools to streamline and reduce the cost of documenting design intent. Lawrence Berkeley National Laboratory created Design Intent Tool (DIT), which provides a structured approach to recording design decisions that impact a facility's performance in areas such as energy efficiency. However, this tool is currently configured for use with hi-tech buildings such as data centres. The Design Intent Tool needs to be reconfigured for common types of commercial buildings and its usefulness tested in a range of situations, and promoted in the industry for widespread use.

2. Automated Functional Testing Tools

Presently there are almost no commercial automated functional testing tools available (the exception is a VAV box tool produced by Siemens). Instead, functional testing is a process that is time-consuming and costly, relying on commissioning providers or skilled technicians who manually test each system and how it interacts with other systems.

LBNL is developing automated and semi-automated versions of a functional testing tool for air handling units that focus on detecting and diagnosing faults in mechanical equipment. Limited testing has been performed; enough to demonstrate proof-of-concept, but there is a need for more comprehensive field testing. There is also a need to broaden the scope of these tools to include testing of controls, other systems and subsystems, including systems in low energy buildings, and operation at the whole-building level.

3. Existing Building Commissioning Cost-Benefit Analysis Tools

Existing building commissioning providers are often faced with a challenging problem. They can identify performance issues, and the necessary corrective measures, but have difficulty quantifying savings without spending a considerable amount of time on calculations or models.

Without these savings figures, they find it difficult to convince the building owner or utility sponsor that the implementation cost is worth the benefit.

Analysis tools are needed to help commissioning providers quickly calculate the energy savings from measures for common and low-energy building systems. Several benefits would result from this streamlined, standardized method of energy savings analysis. At the project level, commissioning providers will be able to provide the information to building owners with data to make implementation decisions, resulting in a higher implementation rate and increased energy savings. The tools will also allow providers to spend less time performing calculations and more time assisting the owner with implementation.

4. Automated Fault Detection and Diagnosis (FDD) Tools

Few commercially available automated FDD tools exist in North America; tools that are commercially available are not yet widely implemented. Prototype tools exist and field testing is underway for specific systems. For example, embedded FDD for AHUs and VAVs is being tested with several controls manufacturers. Demand for these tools is increasing as users see the benefits of continuous monitoring. Current R&D is underway to develop/advance FDD tools, and in the next few years, it will be necessary to extend diagnostic capabilities to additional components that impact operational performance, as well as to the broader, whole-building energy view.

More diagnostic products are needed with tools geared to user needs, specifically:

- Simplify current tools for ease of installation and use by building operators or consultants;
- Define/develop 'dashboards' that link to benchmarks;
- Identify malfunctioning components; and
- Track persistence of savings over time.

4.1.3 Needs from the Japanese Point of View

The term "commissioning" is becoming popular in Japan. Although defined in the SHASE guidelines, the meaning and scope of commissioning varies from engineer to engineer. Recently, energy saving services for existing buildings along with pioneering research work has both been done in the name of commissioning. For these projects many kinds of commissioning tools are used. In most cases computer simulation to estimate performance of HVAC systems or sub-systems and visualization tools are used as a commissioning tool. To date, a variety of FDD tools have been developed in research but their use in real projects is rare for these reasons:

- No commercial tools are available;
- Interface to supply measured data to tools is not provided or inefficient. Users must develop the interface as an extra task;
- Using tools requires too much experience or training; and

- Integrated commissioning tools which can test a variety of HVAC components or sub-systems all at once are not available. This means that tool usage is cumbersome;

The first trial of a functional performance test (FPT) in a real building was planned as a research activity of the SHASE commissioning technical committee (TC) in July 2009 in Japan. This provided an opportunity for a trial use of several commissioning tools to test their performance and determine needs for improvement. Building Systems Commissioning Association is planning a web site to distribute tool resources to volunteer users in collaboration with the SHASE technical committee.

Japanese energy conservation regulations were revised in 2008 and became effective April 2009. The new law requires building owners to report the total energy consumption by summing up all consumptions used in every building owned or rented by each organization. Prior to the revision the report only required reports from building owners whose building energy consumption exceeded a fixed level on the basis of each building. To comply with this revision the installation of measuring systems for energy consumption becomes an inevitable issue, especially in rented buildings and, if a measuring system can only measure the total consumption of a building, a software tool is desirable to estimate subdivided energy consumption per tenant from the total consumption. For example, for an air-conditioning system which serves many tenants and only the total energy consumption is measured, commercial software is available to estimate tenant's energy consumption based on measurements of operating time. Software tools can sum up all energy consumption of a large number of rented areas and provide the summarized information to building users to encourage reducing energy use are going to increase sales as such services rise. This kind of management tool is a very useful tool to rationalize commissioning work and, as a result, helps to expand commissioning.

Energy Service Companies (ESCO) projects are the most common practice for energy conservation in Japan. In ESCO, estimating the baseline of energy use is a key, however, few reliable estimation tools of the nature of a commissioning tool have been developed. There is a need for tool development for application in existing building commissioning as well.

4.1.4 Needs for Existing Building (EB) Cx Tools

Existing building complexity is growing, so new faults occur regularly. In the near future, energy savings will be obtained mainly through optimal control and early fault detection of building HVAC systems as mentioned in Hyvarinen and Karki [23]. Optimization and FDD tools have been seen as promising tools for ongoing commissioning of existing building because of their potential to aid in the persistence of commissioning performance improvements. Many examples of both optimization and FDD tools are found in the literature on existing buildings. Before the needs for existing building Cx tools are discussed, a brief overview of some Cx tools will be given.

Since the existing building energy management control strategies are mostly heuristic, there is a need to systematically examine and improve them, as mentioned in Haung et al. [24]. For example, Lu et al. [25] used a modified genetic optimization algorithm to find the optimal set points of the controllable variables in an HVAC system with cooling coils. In order to obtain effective energy management for an existing HVAC system, an evolutionary programming algorithm was coupled with the simulation tool to provide the optimal combination of the chilled water and supply-air temperatures in Fong et al.[26]. A model based supervisory control strategy, which use a hybrid optimization technique, for online control and operation of building central cooling water systems is presented in Ma et al.[27]. Common for the above tools is that some optimization methods were implemented to find optimal control strategy for HVAC systems, while the methods have not been integrated to BEMS and they have not been widely implemented.

Most FDD tools are based on combinations of predicted building performance and a knowledge-based system. Principles and application of six FDD tools are compared briefly here. A model-based feed-forward control schema for fault detection is described in Salsbury and Diamond [28]. An example of monitoring-based commissioning by use of information monitoring and diagnostics system (IMDS) was reported in Haves et al. [29] and Piette et al.[30]. FDD tools can use the statistical classifier, as reported in the following methods: principal component analysis (PCA) method for sensors as presented by Wang and Xiao[31], the combination of model-based FDD (MBFDD) method with the support vector machine (SVM) method as presented by Liang and Du[32], and the transient analysis of residual patterns as presented by Cho et al.[33]. Air handling unit performance assessment rules (APAR) is a fault detection tool based on expert rules [34]. Even though all the above tools are robust in solving problems, they are still not widely implemented.

Katipamula [35 and 36] gives a summary of the situation of FDD in buildings. He finds that most of the studies completed at the time of the publication (2004) were concerned with individual components or systems. In his article he only presents works which deal with compact air-conditioning equipment, heat pumps, compression chillers and air handling units.

Most of the studies deploy simplified physical models (white box) or black box models in the form of regression models, neural networks or ARX models. Detailed white box models are rarely used.

It is mainly classification methods that are used for diagnosis, giving the character of the residuals (difference between the output signals of the model and the real system). The procedures are used to detect faults with characteristics that are known from experiments or simulations. Although most authors (irrespective of the procedures selected) report that the procedure they used was able to detect the faults sought, Katipamula objects that most surveys were only carried out in the laboratory and not in the field.

Moreover, Katipamula arrives at several conclusions:

- Very few products for FDD exist on the market and these are usually specialised (for individual components) and not automated.

- The (automatic) generation of fault threshold values for detecting faults has not yet been studied in depth, but represents an important aspect in automation.
- The calibration of models requires training data on correct operation. To apply the model on a broad basis it is necessary to develop the model either during the manufacture of the components in the factory, or automated during operation (online). There is still a need for research, particularly on the second approach. Deployment in existing buildings is seen as problematic because measuring data for correct operation is required for the calibration of the model. In existing buildings it is much more likely that equipment and components were not being run correctly before an energy audit.
- Most of the studies mentioned work with procedures for fault diagnosis, whereby significant faults arising singly can be detected. Detection of several faults arising at the same time is thereby not possible. There is also a need for research in this area.
- In general, the amount of measured data available in buildings and its quality is low. The development of cost-effective and reliable sensors is therefore an important step to the spread of FDD in buildings. The gradual introduction of open communication standards in building technology is seen as a major opportunity.
- There is not enough information on the cost-benefits relationship of FDD in buildings.

In Katipamula's view there is a great need for research in the field of FDD in buildings, particularly with regard to practical use and automation.

Based on the existing building needs and current research on FDD and building optimization, needs for Cx tools can be classified as follows:

- **Faults in operation**

Faults in operation can appear due to different reasons, for example faulty construction, malfunctioning equipment, incorrectly configured control systems, and inappropriate operating procedures [37]. In addition, due to abnormal physical changes, ageing, or inadequate maintenance of HVAC components, HVAC components easily suffer from complete failure (hard fault) or partial failure (soft fault) [38]. Since buildings are exposed to different conditions and are used by human, many different faults can always appear.

- **Continuous building change**

The trend with existing buildings is to have them refurbished, renovated and extended; consequently they are acquiring new functions. When a building gets a new function, it is necessary to update that function in a Cx tool . Although a new function may require a new Cx tool for both optimization and FDD, in most of the cases the existing Cx tools can treat the function because of its similarity with the existing. Consequently, development of tools that enables common use is needed to adapt with continuous building change.

Change in the existing building use

An existing building can change purpose completely or partially due to different factors (change in building use, change in building tenants, extension, etc.). To ensure the performance of the building, commissioning should be performed at each change. This required a financial and labor investment, specifically when the process is done manually. This issue can be minimized by implementing Cx tools which rationalize the work by performing optimization or FDD work at each occasion of changes or even better continuously. In such case, automation of Cx tools is an valuable feature.

- **Automation of the Cx tools**

Technology is available for greater application of Cx tools, (wireless communication, data management, web-based communication, etc...) and for their automation. R&D of automatic commissioning is increasingly active. When complete automation is achieved, easy implementation and application of Cx tool in day to day operation will be the reality.

- **New technologies**

Development of new technologies in HVAC systems, building construction and communication necessitate the development of new Cx tools. Currently, web-based databases allow for wider use of energy monitoring. For example, development in data management and web-based technologies provide an opportunity for new Cx tools, while existing Cx tools should be adapted accordingly.

4.1.5 Needs for Low Energy Building (LEB) Cx Tools

Low Energy Building (LEB) are charachterized as those buildings equipped with new HVAC or energy systems such as high COP chillers, inverter controlled pumps and fans, photovoltaic cells, co-generation equipments, etc. as well as environmentally enhanced functions that use natural ventilation, high air tightness and thermal insulation, solar heating, day-lighting, ground thermal energy storage, etc. Commissioning work of these non-conventional systems and building functions is difficult when compared to conventional systems and buildings. Work will improve significantly when tools are available to:

1) Estimate system and building performance under given operational conditions, including weather conditions, in order to test that performance meets design requirement;

2) Carry out fdd of sophisticated systems and building functions; and

3) Optimize hvac and energy systems by the use of simulation.

The reason tools are useful in commissioning is not only due to the sophistication of systems and buildings but also due to the practical limitations of optimization using the system itself. An example of this is an air-conditioning system with seasonal energy storage. Cool energy is stored in the ground in winter and discharged from the ground in summer. The efficiency of the system is determined by the flow rate and the temperature set point of the circulated cold water. The optimal values can be obtained by carrying out experiments many times with different values but one experiment requires one year. This means that getting optimal values is impossible in a practical sense; simulating system performance is an effective approach to solve it. Although many ongoing Cx tools for existing buildings also fit LEBs, the utilization of low energy systems and environmental energy sources require additional tools, especially for optimization)

In this report, Cx tools are classified into two categories; visualization tools and automatic tools. Visualization tools display the performance of an HVAC system in order to detect faulty operation or energy loss. In this case the ability of a tool depends on how adequately selected data and a graph type visualize the symptoms of various faults clearly and effectively. Automatic tools are more sophisticated and can detect faults automatically using computer software. Many Cx tools were developed in previous ECBCS activities such as Annex 25, Annex 34 and Annex 40, however, most are currently only available for conventional HVAC systems and do not address relevant advanced systems and system combinations in LEBs.

4.2 Visualization of Data and Results

Building automation systems (BAS) and building energy management systems (BEMS) have been used in buildings for several decades and are especially important for modern facilities with advanced and low energy building systems. BAS measure, process, and monitor huge amounts of data to operate the building and systems more or less properly. Often, the data is only used to signal failures or break-downs of systems or components. Further use of the data to analyze and diagnose the building operation is limited to the lack of analysis methods and tools.

Advanced visualization techniques can be used to display the information that is otherwise hidden in performance data and/or the recorded operation data of buildings and systems, and, thus, are valuable for commissioning of buildings and systems. Different diagrams and plot types display operation patterns, which quickly help to categorize and analyze large amounts of data, their correlation, frequency, as well as other useful statistical information. These operation patterns allow us to evaluate the quality of operation and to identify optimization potential by comparing them with optimal and/or anticipated operation patterns. Furthermore, the optimization potential can be estimated by additional visual or numerical analysis of the deviation between optimal and existing (real) operation patterns.

Visualization is also relevant for transporting data and information from design analyses, such as simulation and modelling, and making this information available in subsequent project phases.

4.2.1 Classification of Visualization Techniques

Depending on the system to be evaluated, the questions to be answered and the time resolution of the measured data or analysis, different visualizations can be used separately or in combination.

Principally, visualization techniques can be used to show:

- Change of variables over time;
- Relations between two or more variables;
- Statistical information / distribution of values;

- Spatial information/distribution;
- Partitioning/percentage share of properties;
- Comparison of scalars (elements of different size); and
- Process / information flow.

Each of these techniques will be explained in Section 4.2.1.1 thru Section 4.2.1.7.

Before being able to visualize data, significant pre-processing is often necessary. The effort and complexity associated with this task can far exceed that of displaying the data. Here is an example presenting typical pre-processing stepsthat might be necessary when starting with raw data from a BAS or BEMS.

1) Plausibility Check (e.g., remove gross outliers)
Change the time resolution of data (e.g., condense 10-minute values to hourly or daily values)
2) Filtering
"Filtering" denotes the creation of a subset of data that satisfies a certain condition (e.g., subset of the measurements of energy consumption below a certain outdoor air temperature or at a certain operation mode). Thus, the behavior of variables under certain boundary conditions can be studied.
3) Grouping
Data can be grouped according to certain conditions (e.g., heating energy can be grouped for workdays and weekends). Different operation modes can thus be compared.

Filtering and grouping are particularly important to extracting information from measured data, e.g., to show average supply temperature of a heating circuit working independently of the pump control signal.

Additional pre-processing is frequently performed for performance analysis.

4) Data Conversion or Creation
Converting measured raw data into a value with a different or specific unit is often useful for analysis. For example, converting the unit of "electrical energy consumption in kWh" into "energy consumption per unit area in kWh/m2", or "fossil fuel consumed in whole building in kJ", or "emission of global warming gases per year in kg-CO2/year" is often performed. Creating a new value, such as the coefficient of performance (COP) of a chiller or the WTF/ATF value of an air/water distributing system, using the relevant data, is also needed for performance analysis of a component or a system.
5) Calculating Residual Values
Using a new variable by taking the difference between two measured values often enables effective analysis rather than using raw data. For example, differences between inlet and outlet temperatures can show the performance of a chiller or an air-handling coil. The difference between a measured value and the control set point of the value is effectively used for FDD because the magnitude of the difference can easily show a faulty operation.

Selecting the variables to be used in a visualization tool is a critical issue. If an inappropriate set of variables are selected, useful information will not be extracted to detect faulty operation or evaluate performance.

Additional aspects to consider in design of a visualization tool include:

- Tool Aim

  In some tools, measured variables are shown without clear objectives and the tools have no effective use. The purpose of a visualization tool must be as clear as possible.

- Tool Users

  The type of visualization technique used depends on the end user. Current visualization tools are designed to be used by maintenance staff. However HVAC system designers, construction engineers, building owners, tenants, etc. should be involved in actions to increase energy performance. For example, showing energy costs visually to managers is important because they are responsible for budget planning. Appropriate visualization tools for all stakeholders are crucial.

- Software to Build Tools

  Commercial spreadsheet software is used often in data handling and graph drawing. Although ease of use is a great advantage of this kind of software, limited capability in graph design and data handling is a problem. Use of a sophisticated data base system like SQL and specialized graph drawing software should be considered in tool development. This issue is discussed extensively in final report 4 of Annex 47.

4.2.1.1 Change of variables over time

Conventional time series plot

Time series plots are probably the most typical way to display data. The plots in Figure 4-1 show the change of one or more variable over time. As shown in the example, plots can also be stacked.

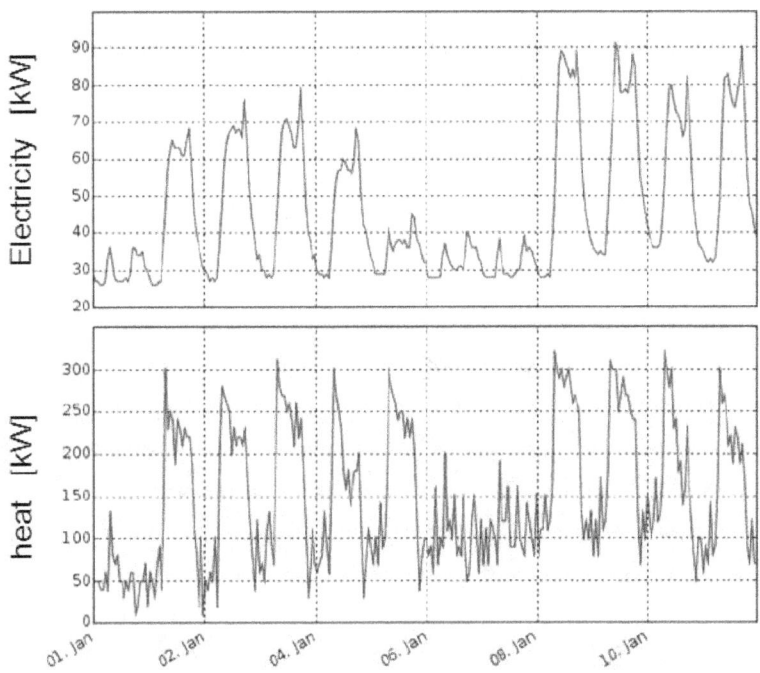

Figure 4-1 Example of a conventional time series plot

These plots are used to evaluate trends, profiles or step-like changes of the variable in time. A significant amount of pre-processing might be necessary if the time resolution of the raw data differs from the time resolution for the desired display.

However, time series plots do not easily reveal the interrelation between two or more sensors in the displayed data.

Carpet plot

Carpet plots are used to display (long) time series of a single variable in the form of a color map, which often reveals a pattern (like weekly operations). Measurement values are portrayed in different colors. For days with a similar course of measurement values the color pattern is respectively similar. Such patterns can be visually identified quickly. This kind of plots helps to identify occupancy and operation schedules. Examples are shown in Figures 4-2 and 4-3.

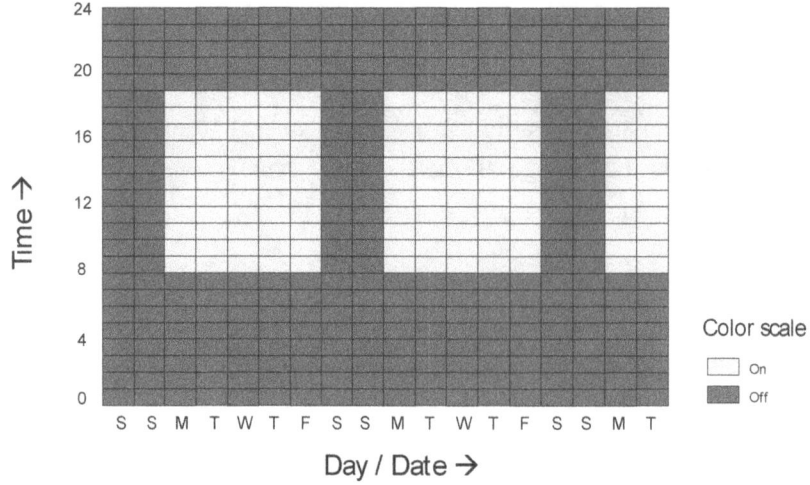

Figure 4-2   **Ideal carpet plot example (e.g., a fan running Monday to Friday from 8.00 to 18.00.)**

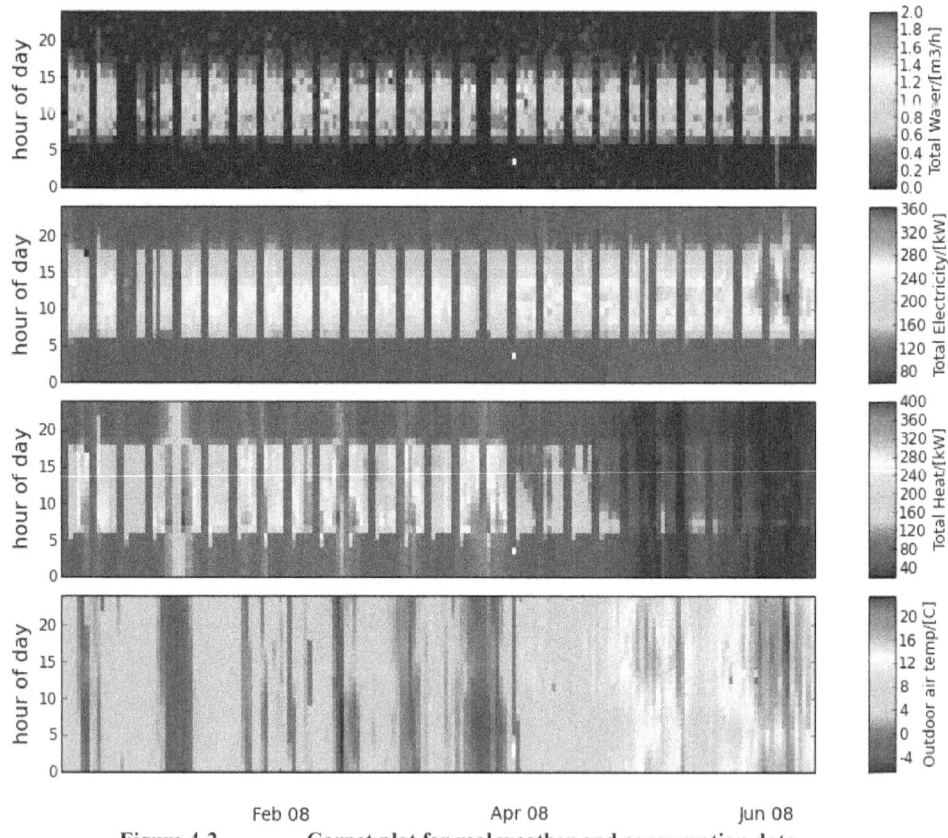

Figure 4-2   **Carpet plot for real weather and consumption data.**

Figure 4-2 depicts and ideal carpet plot where the course of each day runs along the y-axis from the "bottom" (y=0:00) to the "top" (y=24:00) and days are plotted next to each other accordingly on the x-axis. The time resolution is 1 h. Figure 4-3 plot depicts carpet plots of real data. From top to bottom, the plots display water consumption, electrical use, heating, and outdoor air temperatures. Naturally, the patterns are more blurred than in the ideal example.

Layer graph

This type of graph, shown in Figure 4-4, represents a time-series progression from changes in the thickness of each layer, with the objective of showing transitions in the total amount and breakdown of each item. Time is plotted on the horizontal axis, measurements on the vertical axis.

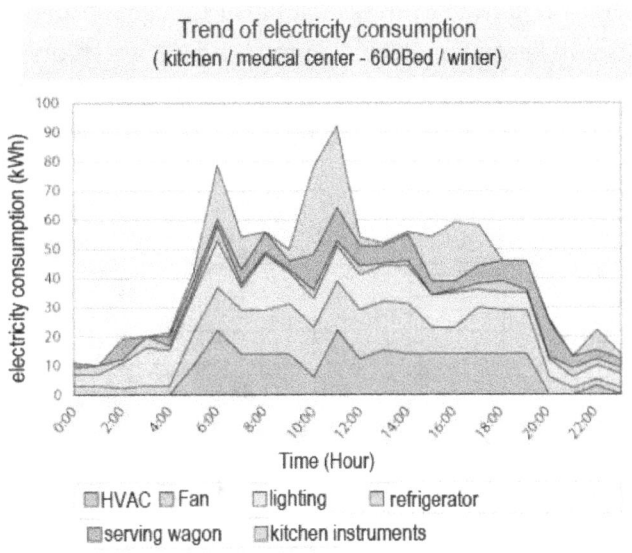

Figure 4-3 Example of layer graph

4.2.1.2 Relations between two or more variables

Correlation plots can be used for the following purposes:

1. To show correlation between controlled and controllable parameters (e.g., to show correlation between outdoor temperature and desired or achieved supply air temperature, CO_2 level and air amount, etc.);

2. For energy auditing of one or many buildings (e.g., to show correlation between the yearly consumption and the floor area);

3. To present a scatter diagram (e.g., the manufacturer's catalogue values can be compared against measured data enhanced by regression lines and presented as the correlation plot).

Scatter plots

This type of graph plots values for items that are thought to be mutually related as distributions along the axes.

Scatter plots show the dependency of two or more variables. Additional information can be gained if the values are grouped (e.g., grouping by day or by a certain operation mode). Potentially, several scatter plots can be combined to scatter plot matrices to show the interdependency of more than two variables.

With the bottom-up approach, this is often used for clarifying details about the subjects being compared, such as the manufacturer's catalog values. A typical usage in a top-down approach would be to generate so called energy signatures (energy demand plotted against outdoor air temperature)

With the addition of regression lines, it can also be used in the analysis of trends or the development of simulation models. Figure 4-5 shows plots that were used to help identify weather dependent control strategies. The time resolution for these plots is 1 day. Each dot represents a daily mean value.The plots are also called a signature or in the case of energy plots, an energy signature.

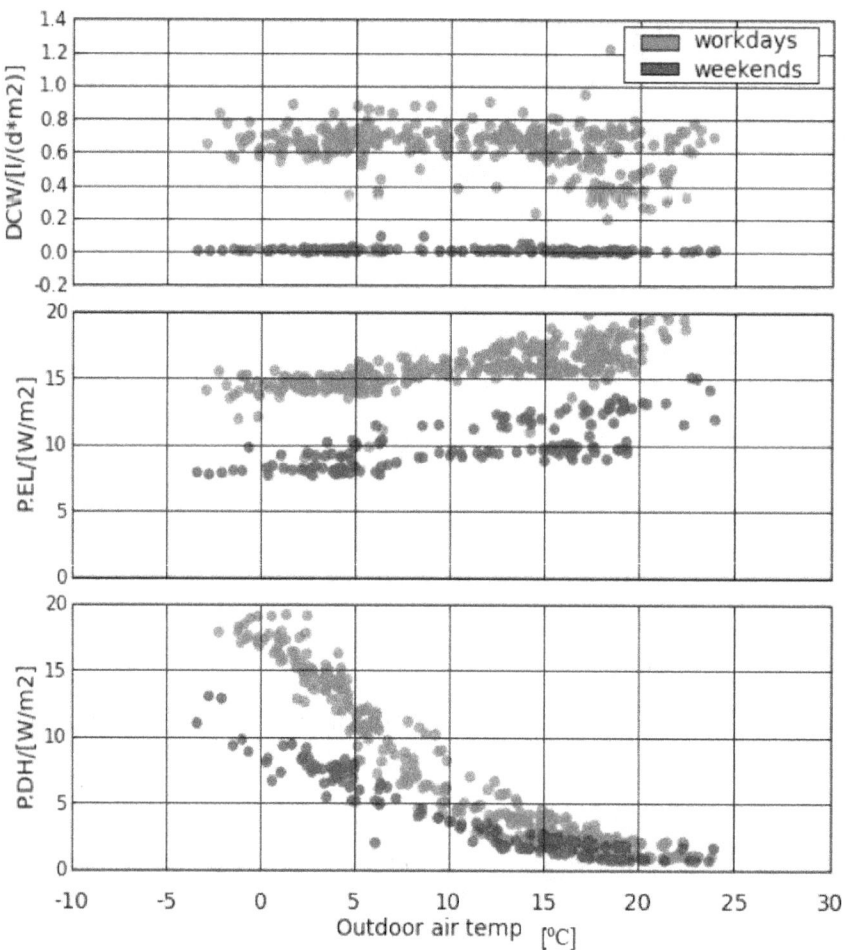

Figure 4-4 Scatter plot of energy and water consumption versus outdoor air temperature, grouped by workday and weekends. [DCW=domestic cold water , P.EL= electrical power, and P.DH=power for domestic heating]

40

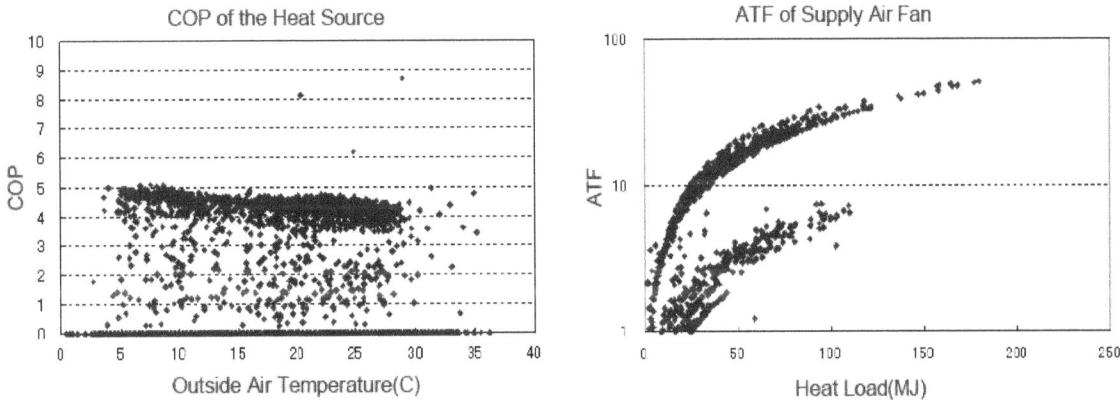

Figure 4-5 On the left, a scatter plot of COP distribution by change of outdoor air temperature, and on the right a scatter plot of ATF distribution of heat load change. [COP is Coefficient of Performance of a refrigerator and ATF is Air Transfer Factor.]

Figure 4-6 shows two plots used to help detect failures or performance degradation.

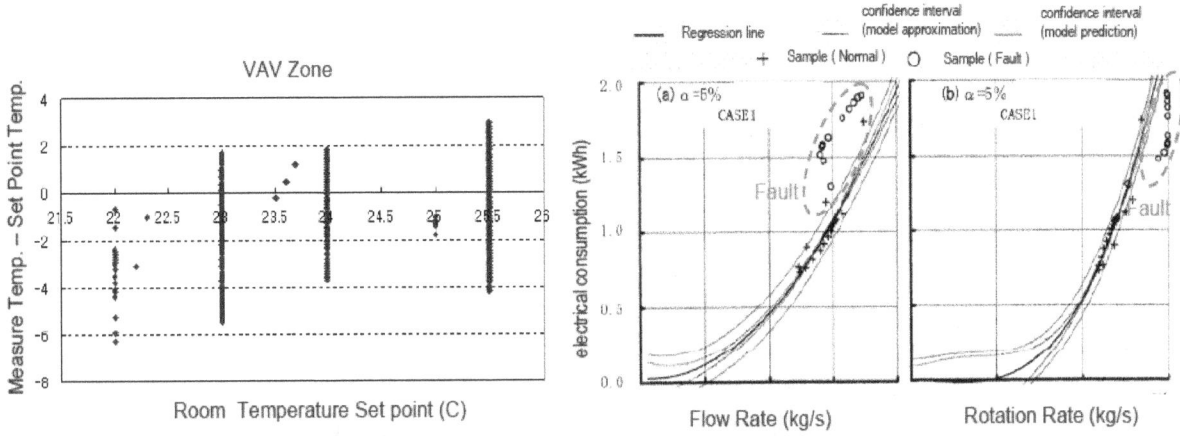

Figure 4-6 On the left, the scatter plot of control error by set value of room temperature. On the right, the right, a scatter plot of the regression analysis results for energy consumption of a supply air fan.

The plot on the left side of Figure 4-7 is used to helps identify failures or performance degradation which appear as outliers. For the plot on the right side of Figure 4-7, the thinner curves indicate confidence range of a fault. Faulty conditions can be statistically detected by the existence of operation points plotted outside the confidence range curves.

4.2.1.3 Statistical information/distribution of values

Histograms

A histogram is one of statistical graphs with frequency on the horizontal axis and series on the vertical axis, which visually represents the frequency distribution. An example is shown in Figure 4-8.

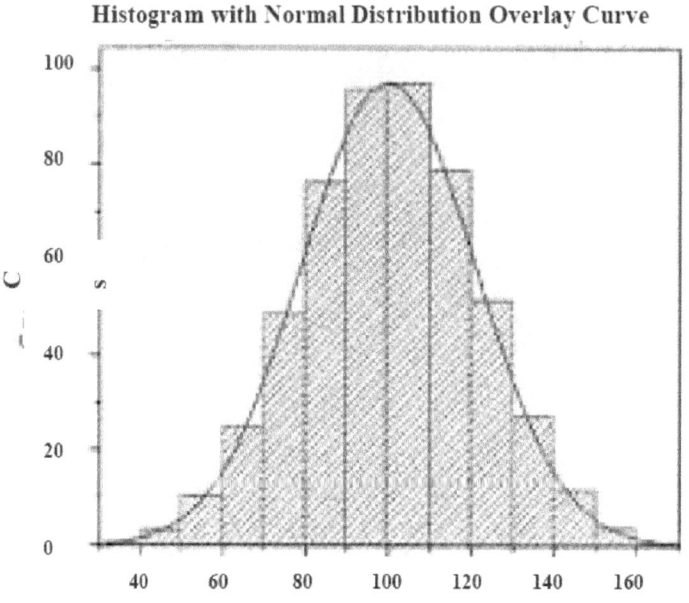

Figure 4-7 Example of a histogram

Descending-order graph

This type of graph, shown in Figure 4-9, plots data for an item in order of size. This is often used for confirming trends in periodic measurements for both bottom-up and Top-down approaches. With a single graph, it's possible to comprehend time-oriented trends or perform annual comparisons. Figure 4-9

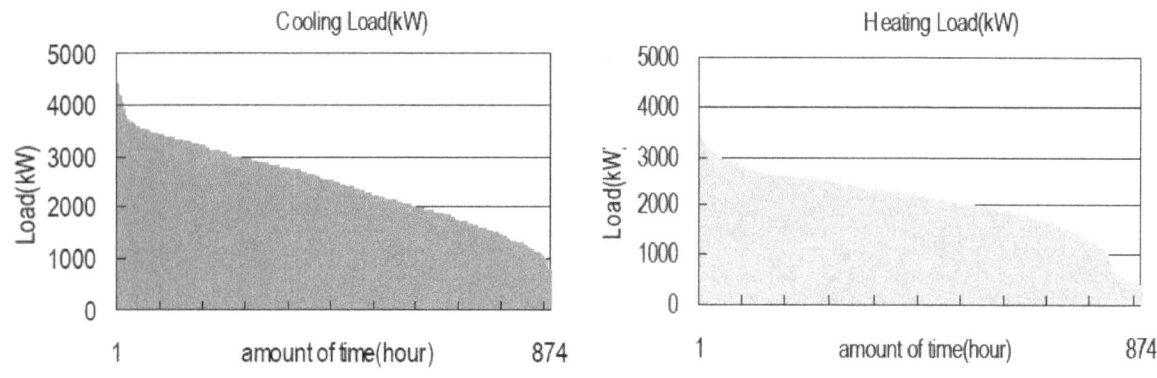

Figure 4-8 Example of a descending order graph

42

Portfolio

This type of graph shows growth rate on the vertical axis and share on the horizontal axis. See Figure 4-10. The circles represent result values. This enables one to clarify the growth level, importance, and location of the point.

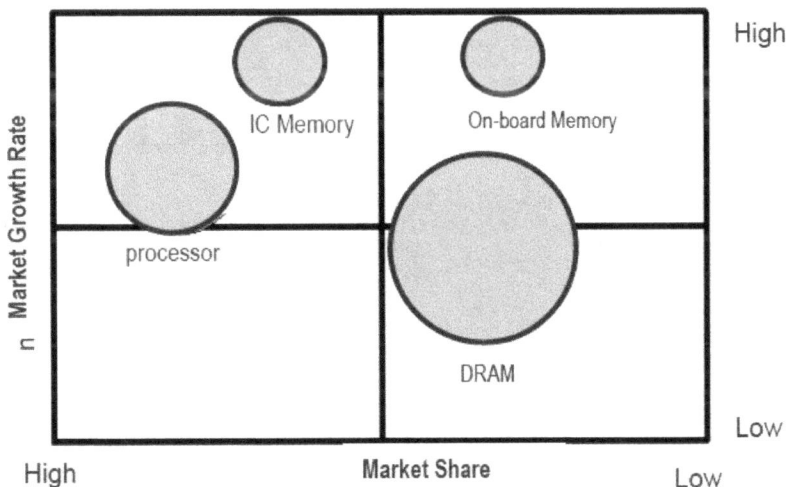

Figure 4-9 Example of a portfolio graph

Box-plot

This graph type presents statistical data and confirms trends.

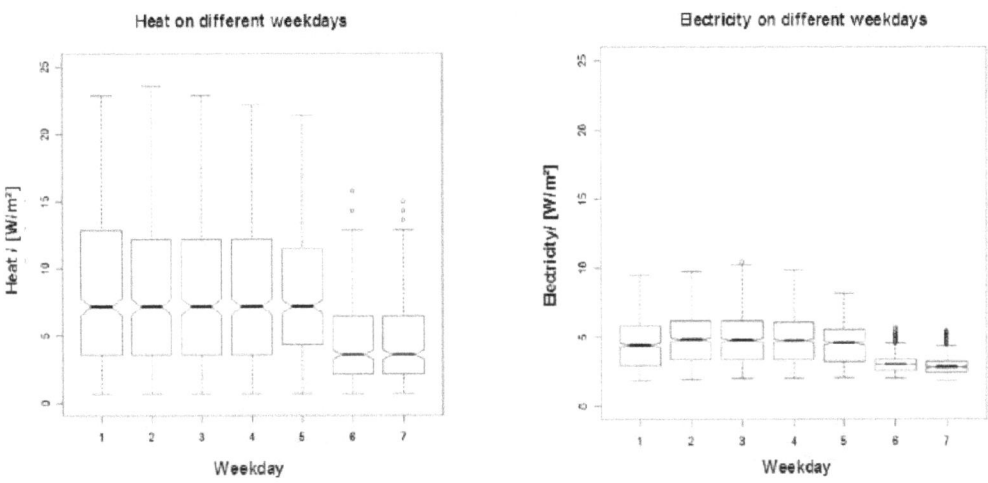

Figure 4-10. Box plots on daily basis

Boxplots display differences between populations. The spacings between the different parts of the box displays the degree of dispersion (spread) and skewness in the data, and identify outliers. They are non-parametric. Figure 4-11 displays heat and electricity consumption on different weekdays. The boxplots shows the difference in consumption between workdays and weekends and the daily distribution.

4.2.1.4 Spatial information/distribution

Maps

Metrics on a map as shown in Figure 4-12 help to understand spatial distribution of statistical values.

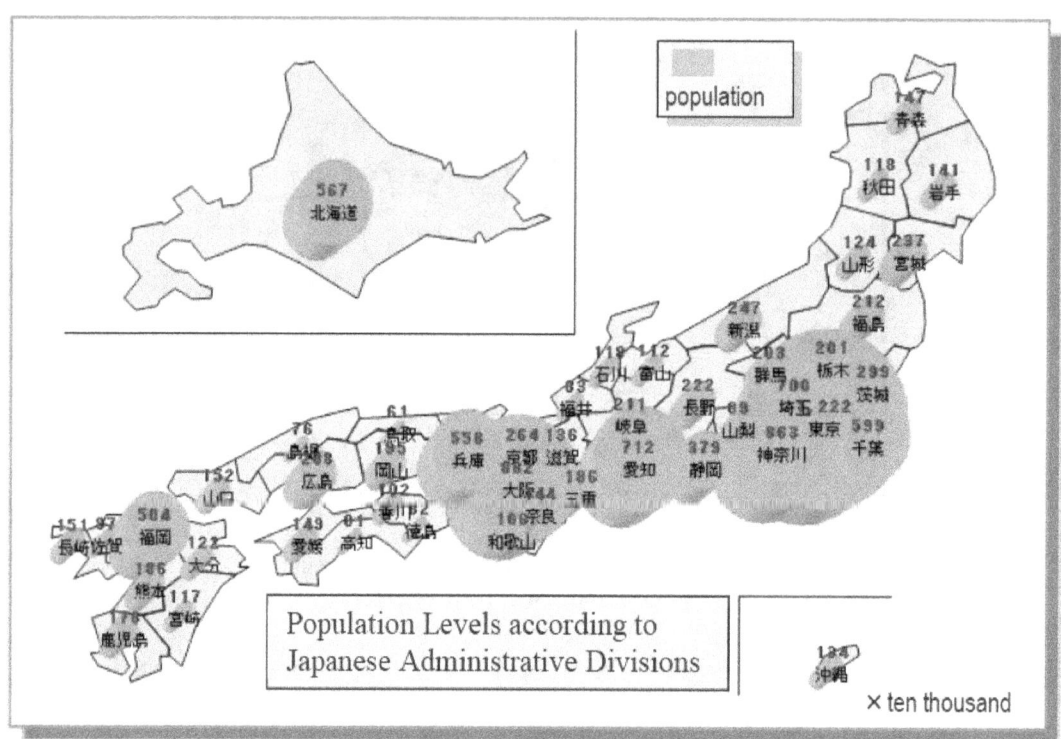

Figure 4-11 Example of a map-based data representation

Pictures (e.g., infrared photographs)

Pictures such as that seen in Figure 4-13 can show heat distribution image by analyzing infrared rays radiated from an object.

Figure 4-12 Example of an infrared photograph

Three-dimentional (3-D) images/ visualization

3D images can show environmental characteristics schematically. The images in Figure 4-14 are examples of the results of CFD analysis representing wind direction, strength of airflow, and temperature distribution by colors and arrows (vectors).

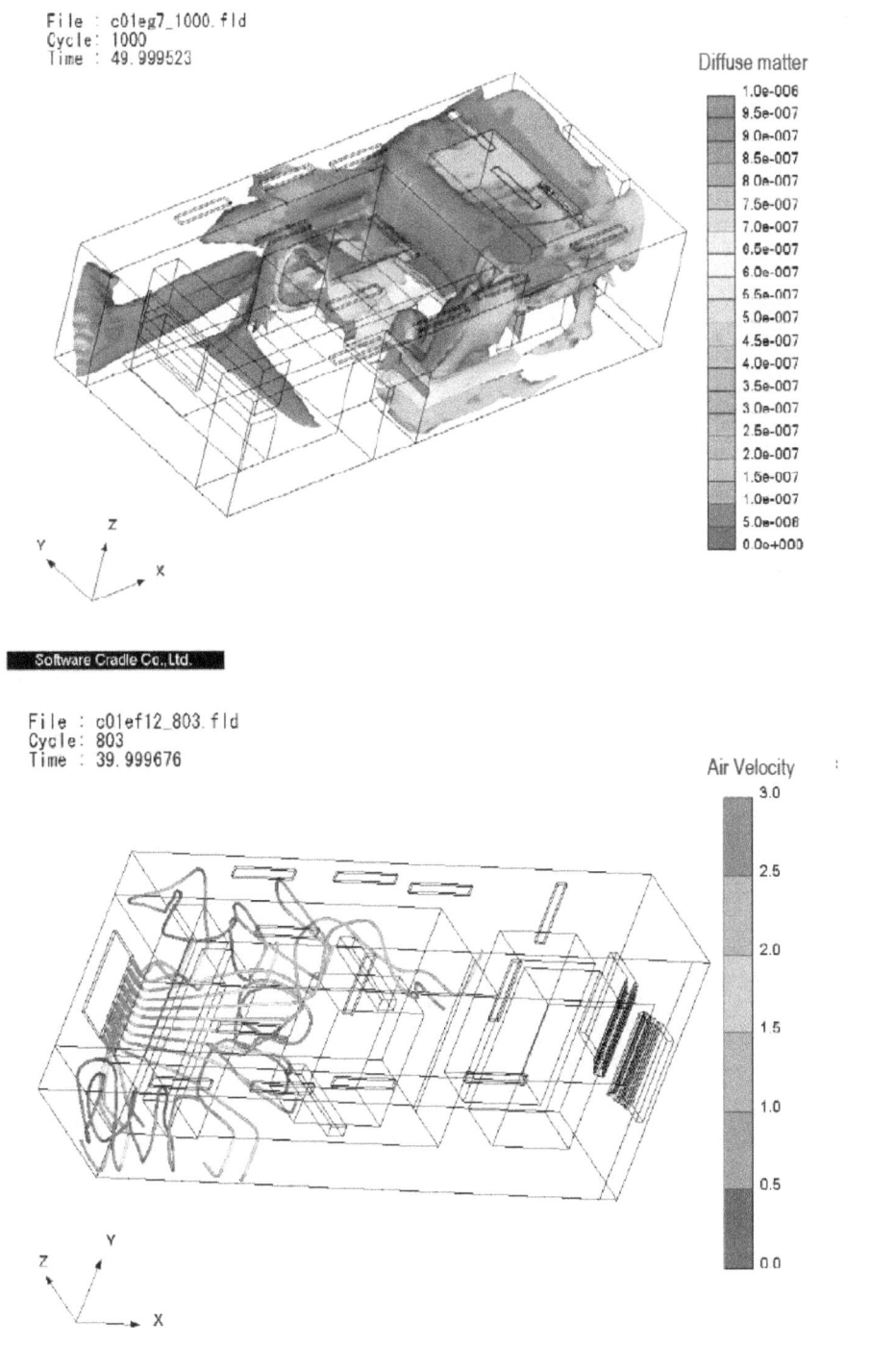

Figure 4-13 Examples of 3-D images

Contour plot

Contour maps elegantly show data attributes and distribution. The example in the upper part of the figure shows changes in air velocity distribution for a particular cross-section. The red dots indicate the position of the temperature sensors used for gathering the temperature measurements on which the CFD analysis was based. The example in the middle part of the figure shows a dimensionless view of the amount of time that air from the inlet reaches a certain location (referred to as "age of air"). The example in the bottom of the figure is a contour representation of the temperature distribution inside a rack-mount unit of a data center.

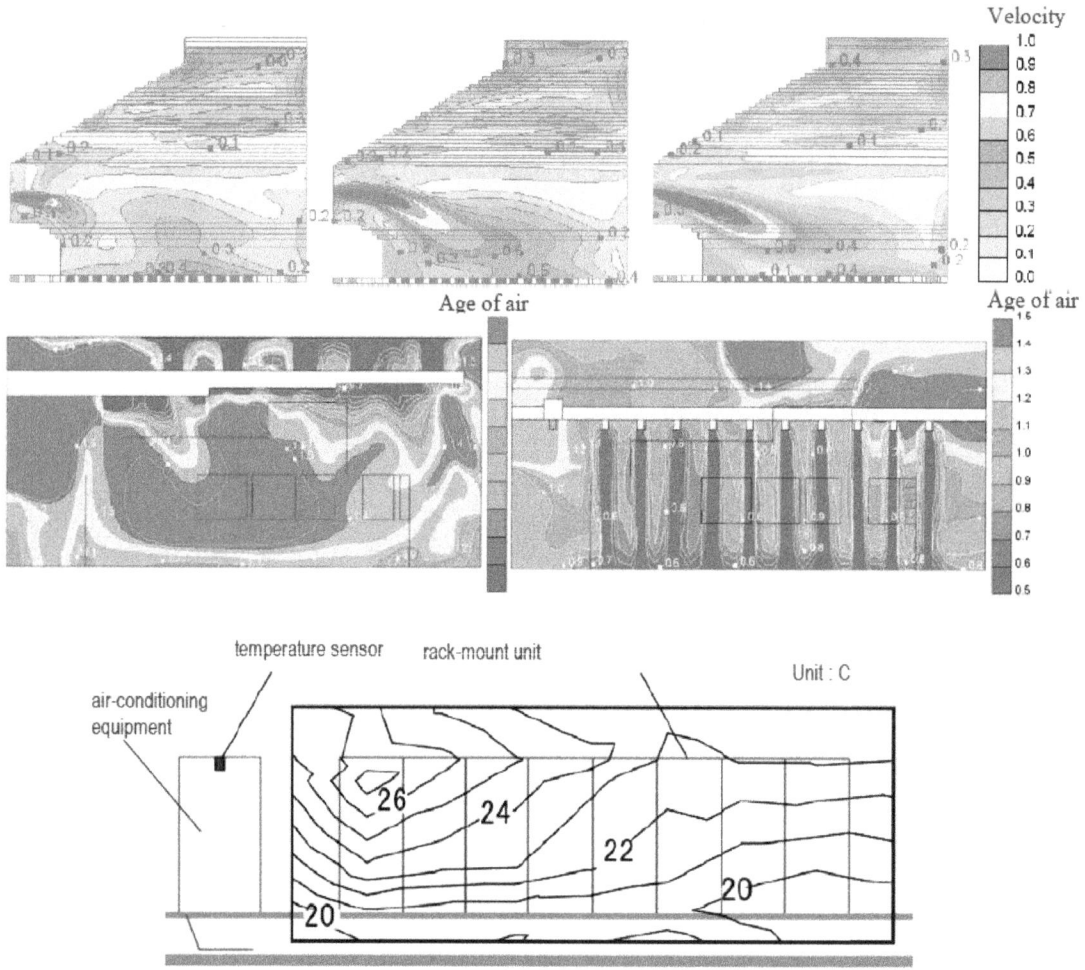

Figure 4-14 Examples of a contour plot

4.2.1.5 Partitioning/percentage share of properties

Pie chart

Pie charts show proportions of each value of the total in a circle format. An example is shown in Figure 4-16.

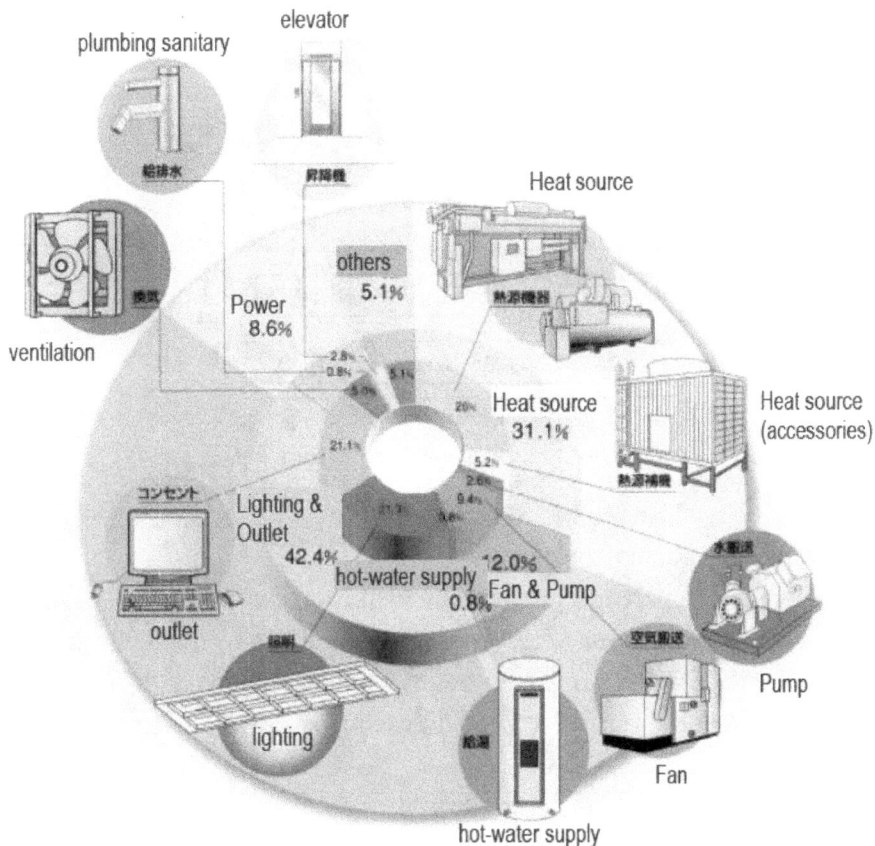

Figure 4-15 Example of a pie chart

100 % stacked bar chart

This chart type is also used to show the proportion of each value. Two examples are shown in Figure 4-17. Proportions are often plotted along the horizontal axis, with the items being compared along the vertical axis.

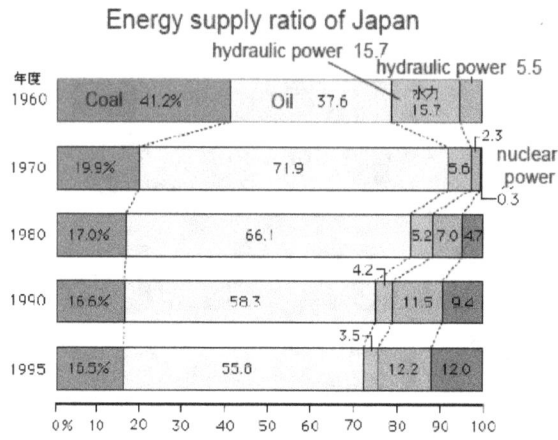

Figure 4-16 Example of a stacked bar chart

Stacked bars (evaluation. normalized)

Stack bars enable comparisons of structural elements, characteristics, or transitions of the items being compared, because the quantities of each part of the items are piled on top of each other. The items to be compared are often plotted along the horizontal axis, with measurements along the vertical axis. An example is shown in Figure 4-18.

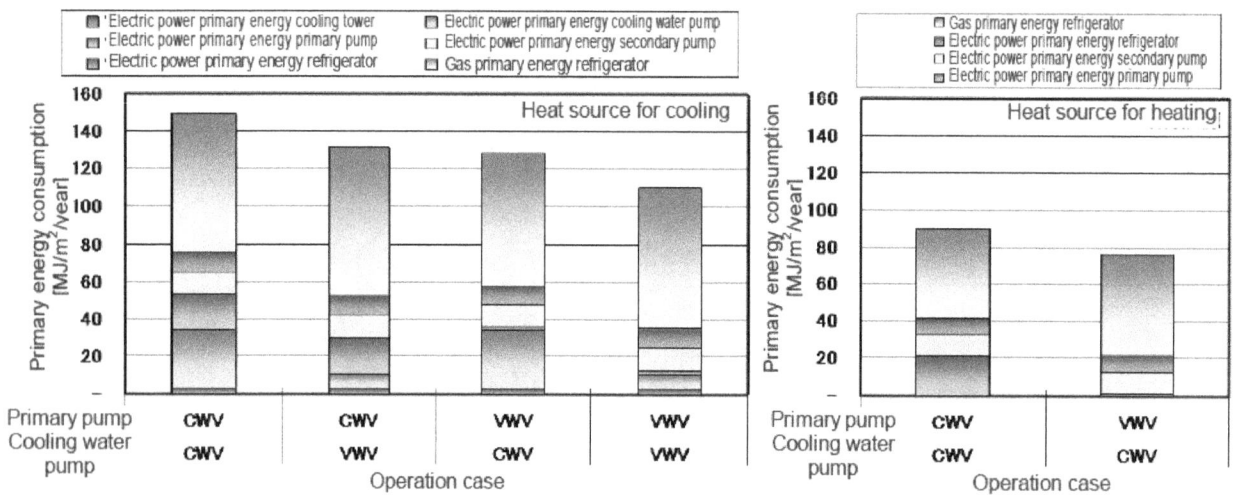

Figure 4-17 Example of a normalized stacked bar chart

Clustered bar chart

Clustered bar chart enables comparisons between items and of overall quantities, with a number of data points associated to each item. An example is shown in Figure 4-19.

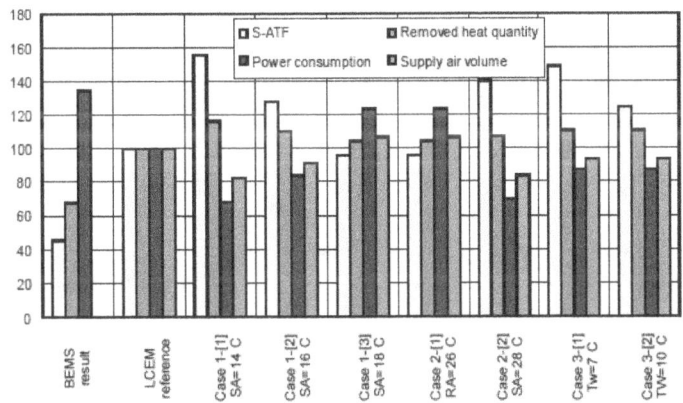

Figure 4-18 Example of a clustered bar chart

4.2.1.6 Compare scalars (elements of different size)

Bar chart

Bar charts compare magnitudes of simple values, and trends over time such as by month. The items to be compared are often plotted along the horizontal axis, with measurements along the vertical axis. An example is shown in Figure 4-20.

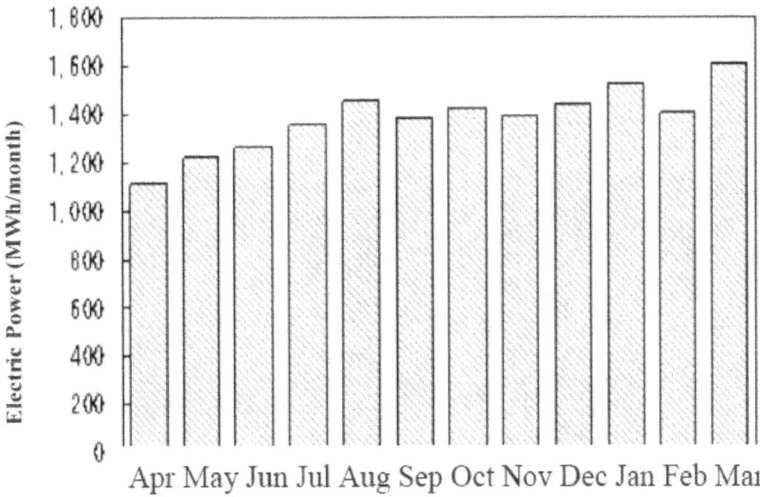

Figure 4-19 Example of a bar chart

Radar chart

A **radar chart** is used to display multivariate data. Three or more variables are represented on axes starting from the same point in the chart. This type of graph is designed to compare the magnitude or quantity of a number of items in a comprehensive manner. An example is shown in Figure 4-21.

49

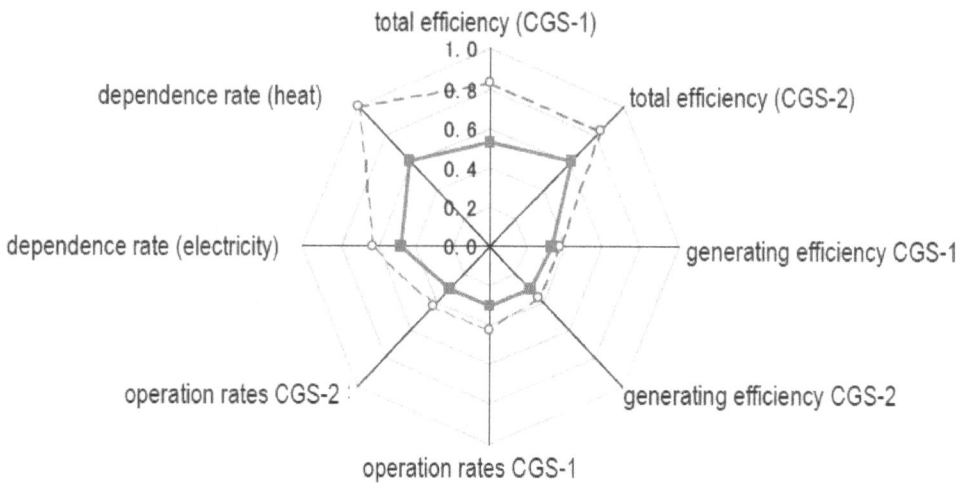

Figure 4-20 Example of a radar chart used to compare the performance of the two co-generation systems (CGS) visually using the targeted values.

4.2.2

4.2.3 Example of Performance Evaluation Process Using Visual Graphs

Data Management

It is advisable to organize the applications of commissioning tools and evaluation indices according to the actual commissioning procedure and the parts to be studied, such as in Figure 4-22, and to clarify evaluation references and other criteria beforehand. An example of energy flows and measurement design in the commissioning procedure is shown in Figure 4-23.

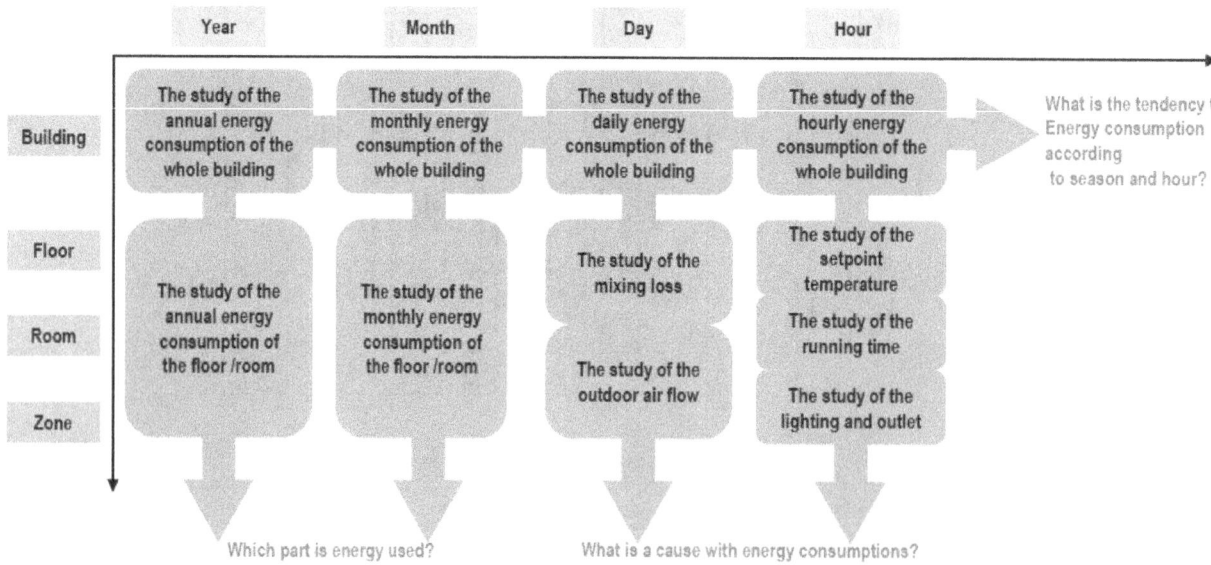

Figure 4-21 Example of commissioning procedure

Classification of entire system, and by energy type

-Measure the energy consumption of the entire building in units of hours or days.
-Understanding the energy consumption state of the entire building makes it possible to optimize energy contracts.
-Analyze consumption trends for each season, day of the week, and time band, and select the optimal energy contract.
-The reporting requirements of the Rationalization in Energy Use Law do not apply at this level.

Classification by energy application

-Determine the energy consumption for each application.
-Understanding the energy consumption for each application makes it possible to select effective energy-saving measures.
-Compare with other buildings and select appropriate energy-saving measures to suit the utilization of the building, such as: if the energy consumption of lighting is high, exchange the lighting stabilizers; or if the transfer power is high, convert fans to inverters.

Classification by energy system

-Determine the energy consumption of each system.
-This makes it possible to discover problem areas and select locations to implement energy-saving measures.
-Once some areas for energy-saving measures have been selected, whether or not the measures really are effective can be assessed by determining the state for each system.
-The system to be dealt with first can be identified by comparing a number of systems.

Specific equipment or floor units

-Measure specific items of equipment or representative floors.
-This makes it possible to decide on update or overhaul, by performing efficiency management on the heat source equipment.
-It is also possible to check behavior in more detail, by focusing measurements on a representative floor, which enables optimization of the operations of facilities and parameter setup.

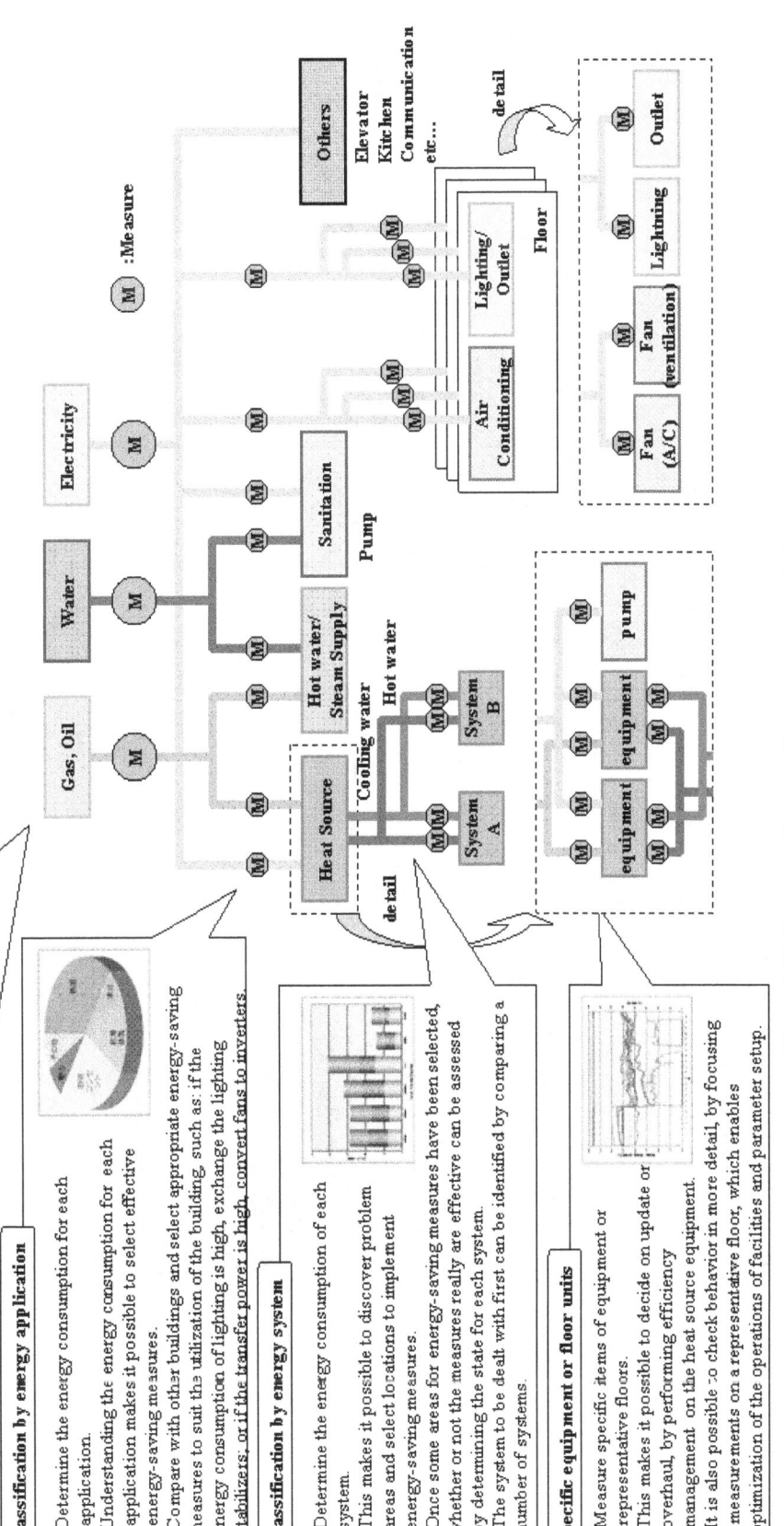

Figure 4-22 Example of energy flow measurement design

51

Example of Performance Evaluation Processes

Tables 4-1 and 4-2 present examples of performance evaluation processes implementing visualization techniques for an entire building, and for building components, respectively.

Table 4-1 Example of an evaluation process for an entire building

Graph	Summary	Indices	Evaluation Method
	• Compare the primary energy consumption with those of other buildings of the same type. • Scatter diagram with total floor area p o e a ong e horizontal axis and equivalent primary ener... consumption along vertical axis. • Remember that the characteristics of the buildings used for comparison are not clear.	• Amount of incoming electrical energy • Amount of gas consumed • Amount of mains water used • Amount of chilled water received (with DHC) • Amount of steam received (with DHC)	• For the same applications, the plots will be close to the regression line. • If the amount of energy consumed is greater than that of other buildings, compare details and detect problem areas.

- Detect whether there is a peak.
- Perform load equalization, with the aim of reducing installed capacity and energy costs.
- Compare daytime and nighttime consumptions for hourly rates, or workday/weekday consumptions for daily rates.
- Understanding the load increase/decrease characteristics will facilitate the progress of energy-saving investigations.

- Amounts of incoming electrical energy, gas, and oil
- Amounts of electrical energy, gas, and oil used in applications

- Determine time bands in which consumption is highest, from the trends in amounts of electrical power (or gas or oil) consumed in the entire building.
- Estimate energy saving and load equalization items from the consumptions for each application.
- Compare load fluctuations by plotting data such as load ratios in a clustered bar chart.

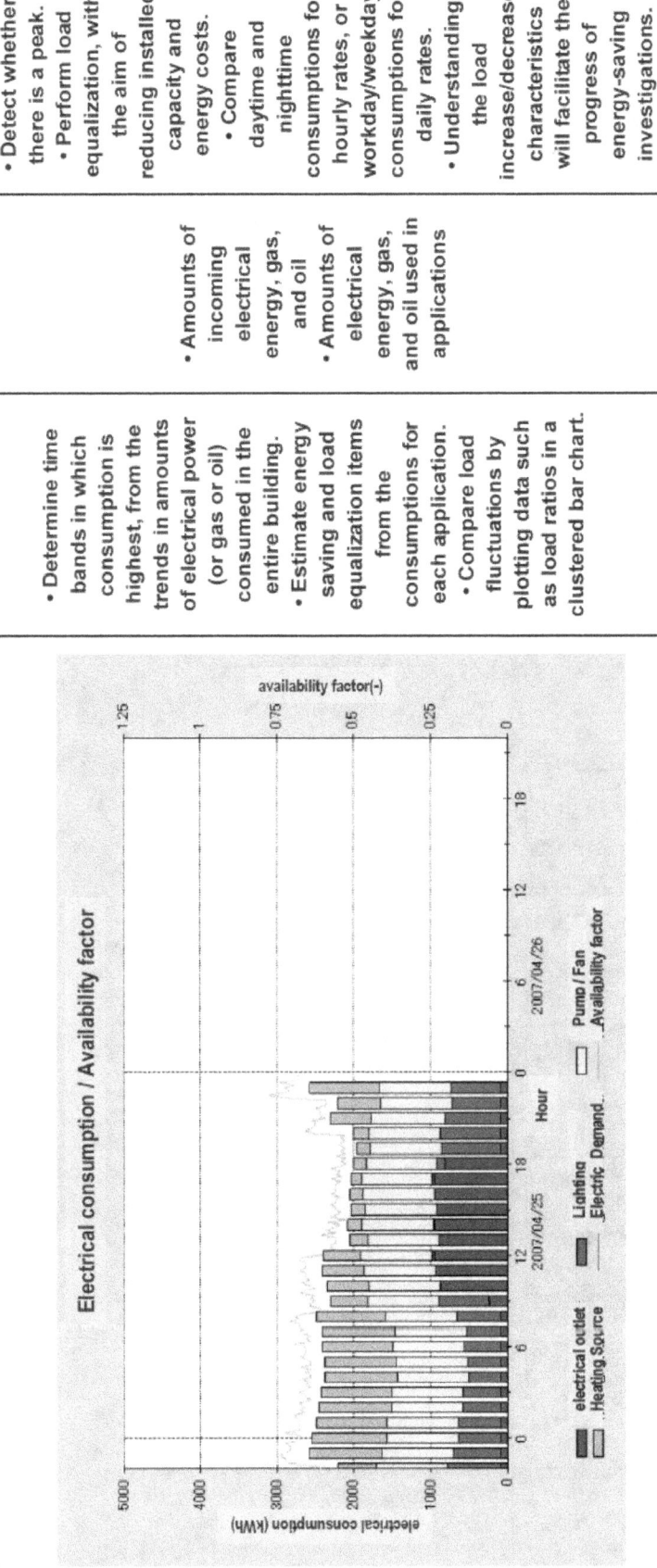

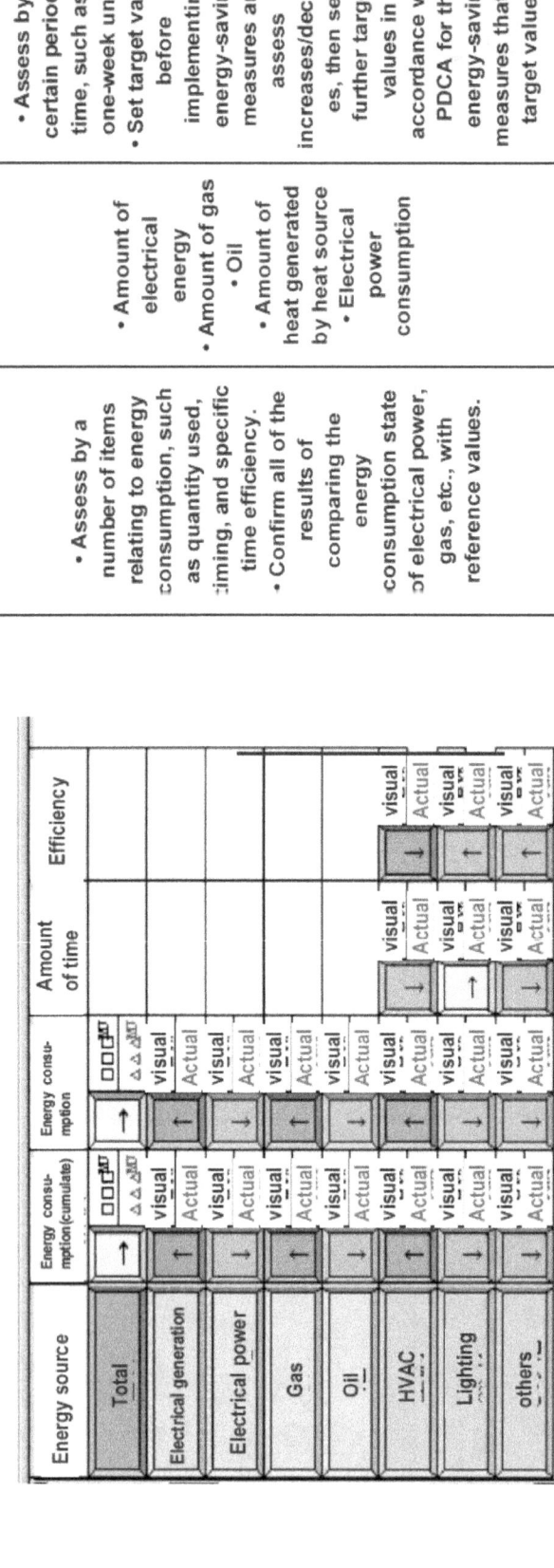

- Assess by a number of items relating to energy consumption, such as quantity used, timing, and specific time efficiency.
- Confirm all of the results of comparing the energy consumption state of electrical power, gas, etc., with reference values.

- Amount of electrical energy
- Amount of gas
- Oil
- Amount of heat generated by heat source
- Electrical power consumption

- Assess by a certain period of time, such as in one-week units.
- Set target values before implementing energy-saving measures and assess increases/decreases, then set further target values in accordance with PDCA for the energy-saving measures that set target values.

54

• Select items for the focus of energy-saving efforts, from the proportions of the amounts of electrical power (gas, oil) consumed by the entire building. • Determine changes in the overall consumptions and changes in the proportions of energy consumptions for the entire building caused by reductions, by comparing data before and after the implementation of energy-saving measures. • Select items to be measured.	• Amounts of incoming electrical energy, gas, and oil for the entire building • Amounts of electrical energy, gas, and oil for each application.	• Make rough estimates of the effects of energy-saving measures from changes in the amounts of energy consumed and energy saving ratios. • Compare proportions by application before and after the implementation of energy-saving measures, and confirm any effects on other applications due to those measures. • Investigate measures aimed at items which always have the largest proportions, to maximize effects of the energy-saving measures.

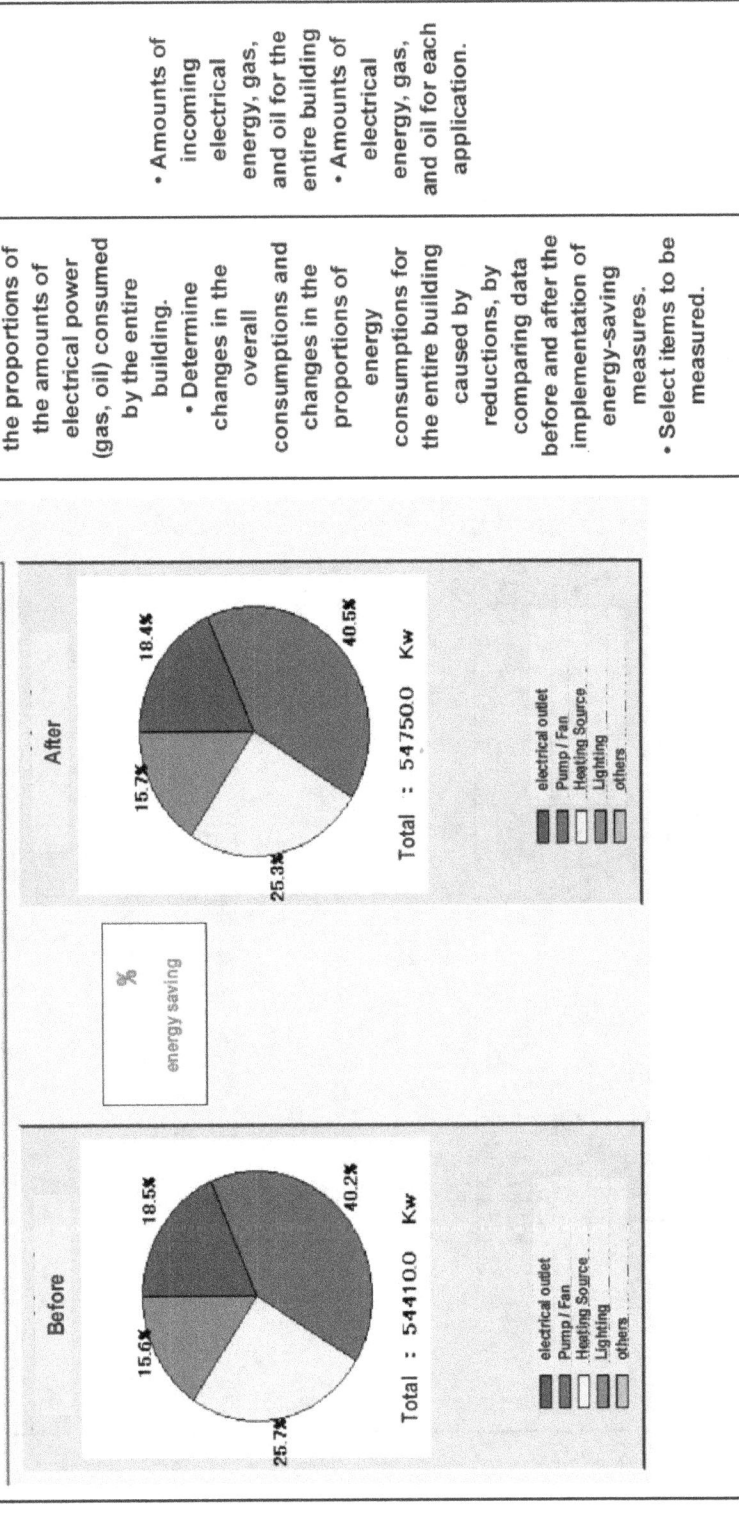

Table 4-2 Example of an evaluation process for building components

Graph	Summary	Indices	Evaluation Method
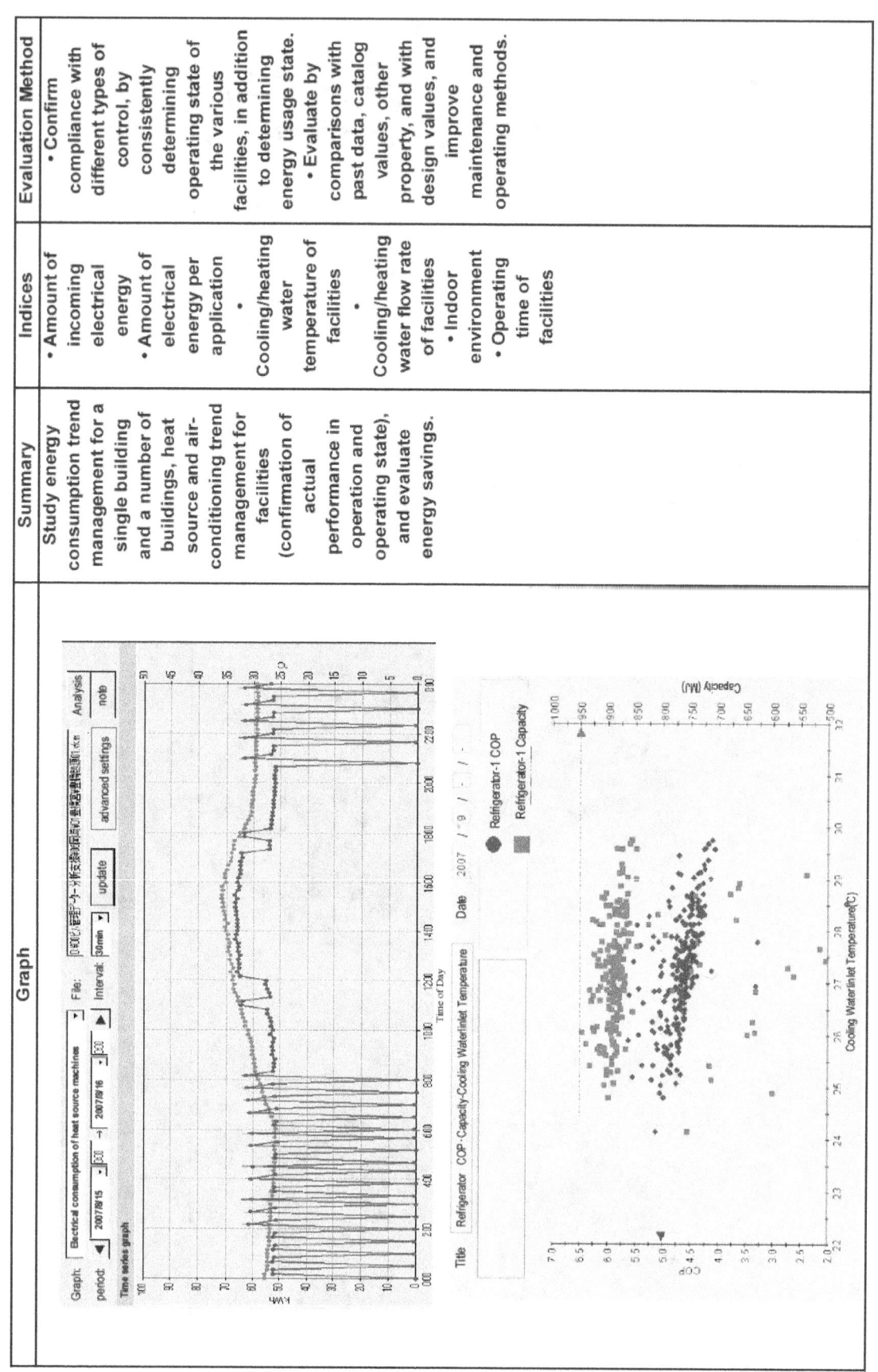	Study energy consumption trend management for a single building and a number of buildings, heat source and air-conditioning trend management for facilities (confirmation of actual performance in operation and operating state), and evaluate energy savings.	• Amount of incoming electrical energy • Amount of electrical energy per application • Cooling/heating water temperature of facilities • Cooling/heating water flow rate of facilities • Indoor environment • Operating time of facilities	• Confirm compliance with different types of control, by consistently determining operating state of the various facilities, in addition to determining energy usage state. • Evaluate by comparisons with past data, catalog values, other property, and with design values, and improve maintenance and operating methods.

• Ensure that it is possible to create graphical displays selected from a large number of target items, so that even managers with little specialist knowledge can extract items with large contribution ratios. • In determining specific operating conditions for efficiency improvement, create a scatter diagram of items with large contribution ratios and chiller COPs, for further study.	• Amount of heat generated and power consumption of each chiller • Chilled water outlet temperature, chilled water inlet temperature, chilled water flow rate, coolant water inlet temperature, chiller evaporator temperature, chiller condenser pressure	• Calculate and compare contribution ratios for items that can be assumed to have causal relationships: chiller COP, chilled water outlet temperature, and coolant water inlet temperature. • Assess if causal relationship exists for each item, by displaying data on a radar chart. • These results can be used when analyzing various causes, by varying objective variables and explanatory variables.	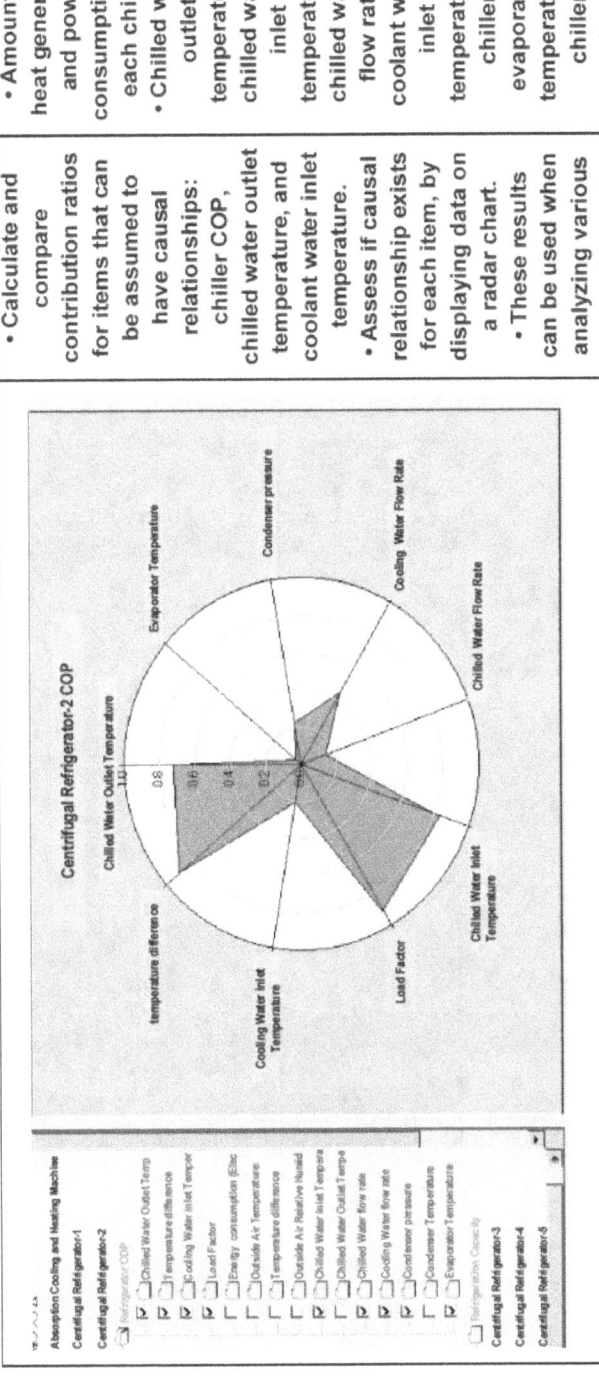
Check that no residual storage occurs. The system operates optimally if the combined ambient temperatures are at design values. If there is residual storage, follow the procedure below to modify the control parameters.	• Amount of heat generated by each equipment item of • Amount of heat remaining in thermal storage tank	• Optimal tuning of heat source equipment of thermal storage tank system • Tuning of heat source equipment to suit actual loading • The points are the start/stop points of the follower operation equipment.	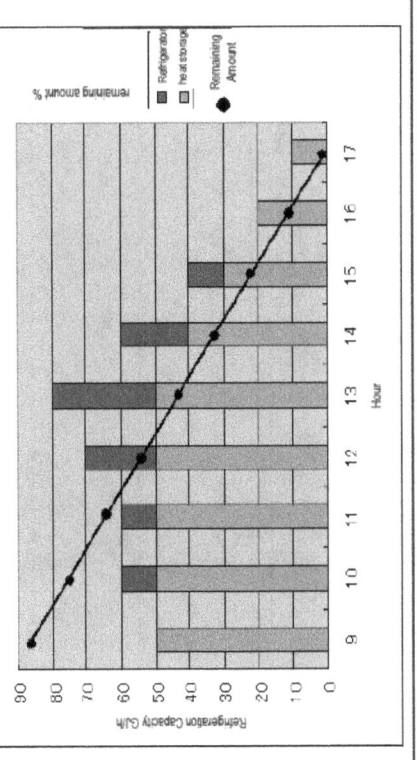

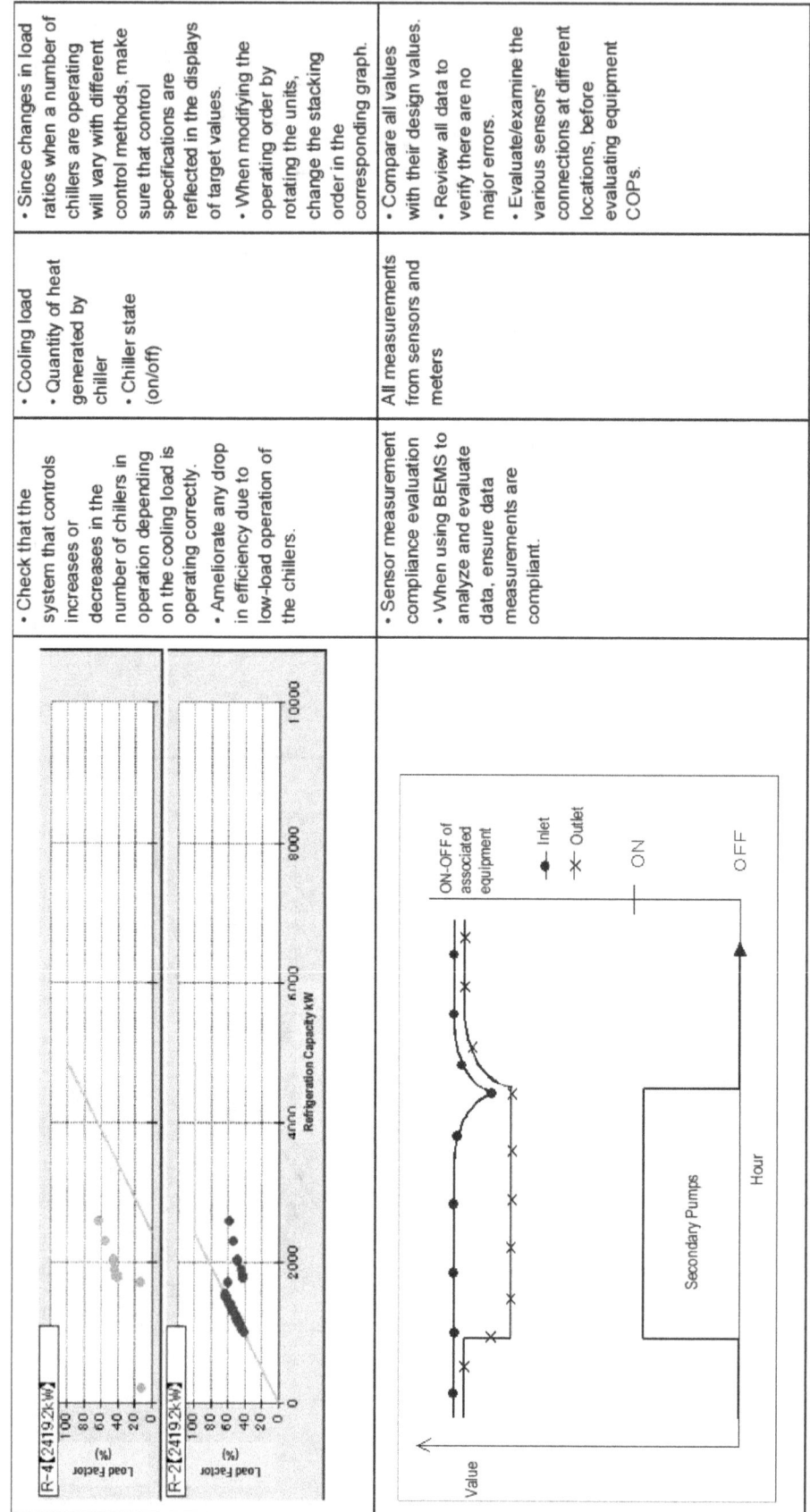

• Check that the system that controls increases or decreases in the number of chillers in operation depending on the cooling load is operating correctly. • Ameliorate any drop in efficiency due to low-load operation of the chillers.	• Cooling load • Quantity of heat generated by chiller • Chiller state (on/off)	• Since changes in load ratios when a number of chillers are operating will vary with different control methods, make sure that control specifications are reflected in the displays of target values. • When modifying the operating order by rotating the units, change the stacking order in the corresponding graph.
• Sensor measurement compliance evaluation • When using BEMS to analyze and evaluate data, ensure data measurements are compliant.	All measurements from sensors and meters	• Compare all values with their design values. • Review all data to verify there are no major errors. • Evaluate/examine the various sensors' connections at different locations, before evaluating equipment COPs.

4.3   Automated Tools

4.3.1   General Features of Existing Building and Low-Energy Building Commissioning

Eight categories for discussion are used when describing features of existing building (EB) and low-energy building (LEB) commissioning tools: objectives, functions, data-management, implementation, operability, analytical engine, end users, and benefits. Each of these categories are discussed here.

Objectives

The objectives of EB and LEB commissioning tools are diverse. However, the main objectives are: estimating design performance, detecting and diagnosing faults (FDD), optimizing HVAC operation without sacrificing room environment, and auditing energy consumption based on statistical or simulated reference values. Tool targets range from whole buildings to components, in addition to control systems. Some sophisticated tools will have multiple objectives, for example, ranging from fault detection to optimal operation. Some tools will have a single objective such as analyzing the performance of a natural ventilation system. Conventional tools could be easily applied to EB with minor modifications; however tools for LEB must incorporate new features.

Section 4.3.2 presents 18 new tools developed in Annex 47. Table 4-3 shows the objectives of each tool. The number of commissioning tools aimed at LEB is low with only one tool existing for LEB commissioning in design review.

Functions

Each commissioning tool has a variety of functions to achieve its objectives. Verification testing using results from simulation or statistical analysis is the basic function of a tool when performing FDD, building optimization and HVAC systems and design review. As seen in Section 4.2 on HVAC systems performance, verifying test results, fault existence, etc. by visual and articulated graphs, charts and tables is another important feature of tools. Although most tools can produce commissioning reports with little labor cost through computer use, this type of auxiliary function is highly valued in real applications because of the importance of budgets.

The functions of a sophisticated tool are often complex. Verbally explaining the functions and structure of such tools with any accuracy is difficult. In this report an IDEF∅ model is used to systematically explain a Cx task's complicated structure. IDEF∅ is a method designed to model the decisions, actions, and activities of an organization or system. IDEF∅ was derived from a well-established graphical language, the Structured Analysis and Design Technique (SADT). Precise explanation of IDEF∅ is found in Report 4 of Annex 47.

Figure 4-24 shows an IDEF∅ model for the tool used to optimize an HVAC system with ground thermal storage system. The left-hand figure shows that the tool estimates the optimal operation values (output) from the point of view of energy conservation. Optimal values are obtained using:

- Weather conditions and room air conditions (input);

- Various set points, operational conditions and system specifications (control); and

- System models and software programs (mechanism).

In general, IDEFØ expresses the parameters including mechanisms and controls, at the top and bottom of each box, respectively. The controls represent the application's related constraints (cost, time, regulations, etc.). Mechanisms are resources (tools, methods, etc.) that enable the application to work.

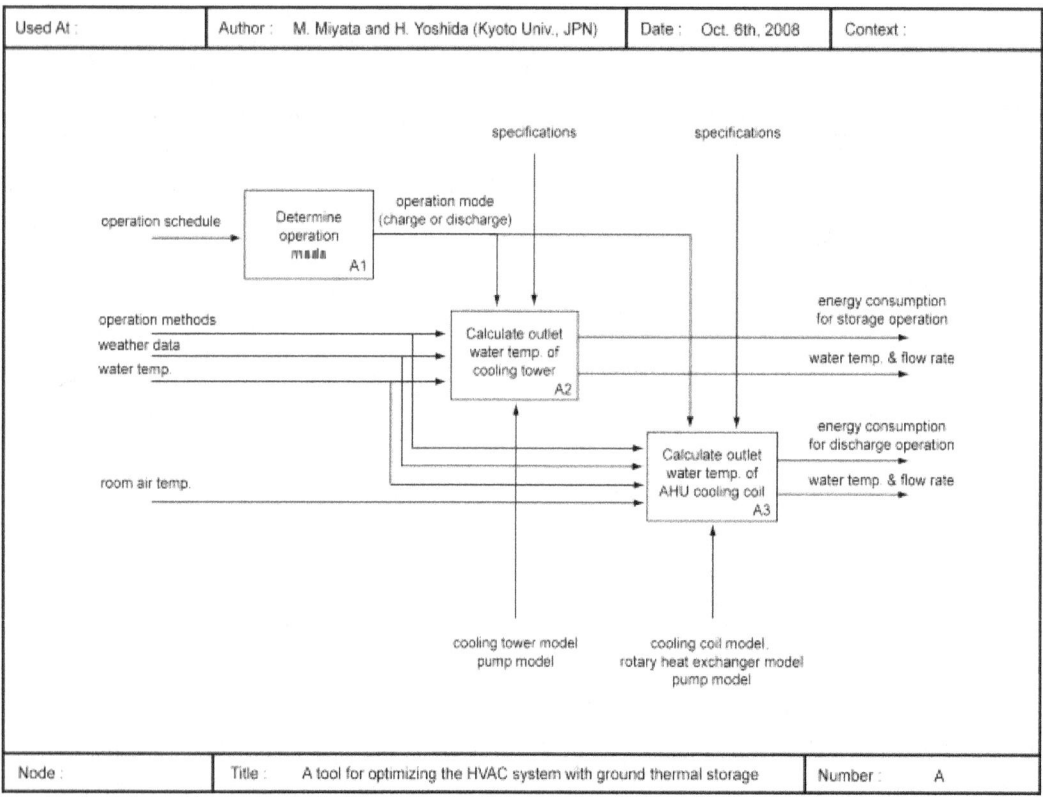

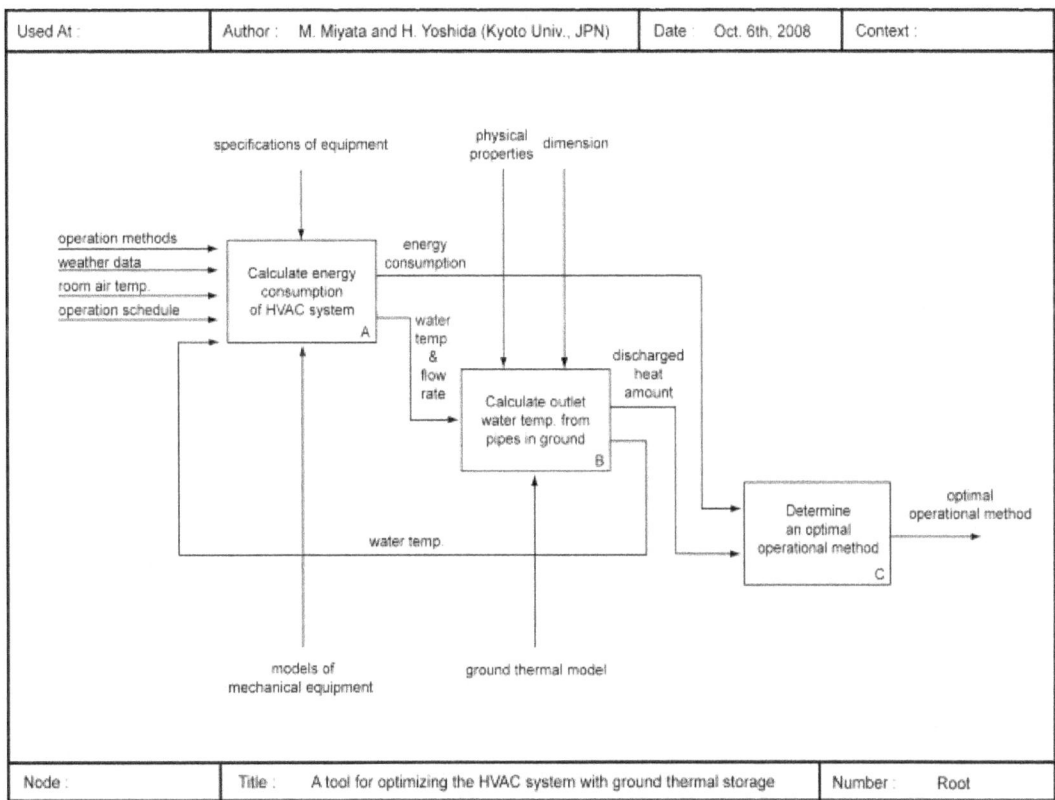

Used At :	Author : M. Miyata and H. Yoshida (Kyoto Univ., JPN)	Date : Oct. 6th. 2008	Context :

specifications of equipment

physical properties   dimension

operation methods
weather data
room air temp.
operation schedule

Calculate energy consumption of HVAC system   A

energy consumption

water temp & flow rate

Calculate outlet water temp. from pipes in ground   B

discharged heat amount

Determine an optimal operational method   C

optimal operational method

water temp.

models of mechanical equipment

ground thermal model

Node :	Title : A tool for optimizing the HVAC system with ground thermal storage	Number : Root

Figure 4-23 Example of IDEFØ

Data Management

A variety of data types are used in a commissioning tool. They are classified into five types: design data indicating thermal performance of building walls and HVAC component performance; data of system state such as chilled water temperatures and flow rates; environmental data such as weather and room air humidity; data representing the operation schedule; and statistical data such as the baseline of monthly energy consumption.

Most tools are designed to obtain all data except design data electronically from BEMS or via the Internet, although in some cases data may be

BIM: Building Information Modeling

STEP: Standard for the Exchange of Product Model Data

IFC: Industry Foundation Classes

XML: Extensible Markup Language

AEX: Automating Equipment Information Exchange

CSV: Comma Separated Values

SQL: Structured Query Language

HDF: Hierarchical Data Format

obtained through off-line data files. (A list of abbreviations for data processing models is listed in the call-out box.) The amount of design data increases with a building's size; as a result, they are stored in a computer aided design (CAD) system database with a data structure known as Building Information Modeling (BIM.) New data structure systems defined by Standard for the Exchange of Product Model Data (STEP), Industry Foundation Classes (IFC), aecExtensible Markup Language (XML), and Automating Equipment Information Exchange (AEX) are becoming the standards. By using this kind of data structure more effective data exchange or data share is achieved compared to a data structure like Common Separated Values (CSV). Usage of an appropriate database like Structured Query Language (SQL) is essential but all the tools developed within Annex 47 do not necessarily have this feature yet. Simpler data structure like Hierarchical Data Format (HDF5), which is not as sophisticated as IFC, are used especially for processing measured data.

In existing building (EB) cases it is difficult to obtain design data because the database is typically inaccessible. Some tools require training data to run. In case of an EB, training data are normally available but for a new low energy building (LEB) they are not. Therefore some tools may not be used with LEB.

Data transfer protocol is another important issue in data management. The protocol used in BACnet and LONTalk, which are open networks, enables effective data exchange. But in many EB the protocol cannot be used due to communication protocols which are proprietary to a specific manufacturer or are not open.

Implementation

Hardware and software issues affect the implementation of tools on computers. Since computer calculation capabilities have increased while the cost of computers and related hardware devices have decreased, hardware issues pose less of a challenge. In contrast, issues related to software management, particularly program language, remain significant. Tools have been and are still being developed using various languages such as FORTRAN, C, JAVA, MATLAB, etc. As a result, developing a new tool that is compatible with existing tools that use various languages is challenging. This is why certain tools are operated semi-automated or are otherwise not fully automated.

Operability

A tool's operability depends on its simplicity and ease of use. Its features rely heavily on performance of the visual interface featured in the tool. Many types of graphic sub-tools are available and suitable visual environments are easily found and implemented. Some tools utilize the visual environment provided for in software functions such as MATLAB/Simlink. Operability of the tools developed in Annex 47 varies; however many unique and convenient visual interface tools were implemented as seen in Section 4.4.

Analytical Engine

A variety of analytical engines are available whose mechanisms are based mainly on simulation, statistical analysis, linear and non-linear programming, expert rules, and fuzzy logic. Many analytical engines were developed and tested in real commissioning projects during Annex 25, Annex 34 and Annex 40. In addition, many analytical engines are also available as commercialized or open source software. For example, common CFD software and air-conditioning load calculation software like EnergyPlus are used as an analytical engine in some tools. For LEB commissioning, however, development of new types of analytical engines which can estimate a building's energy performance using CFD software or heat flow simulation of the building envelope is needed. Some examples are shown in Section 4.4.

End Users

The principal end users of commissioning tools are commissioning providers; however, building owners, building managers, facility managers, operators of HVAC systems are also potential users. Generally speaking, each user requires different functions. For example, building owners and managers are interested in tools that help reduce energy consumption, while commissioning providers and HVAC system operators want tools that help find and diagnose faults for system optimization. If building system designers use some tools at the planning or designing phase and store the information prepared for the tool in a data base, quality, productivity and cost of commissioning work will improve because seamless information flow is established throughout all construction phases. This is what Annex 47 emphasizes for future commissioning processes.

Benefits

Commissioning tools provide many benefits; however, the major benefit is energy conservation, especially in EB. Other benefits are improved room environment and a resulting reduction in resident complaints, increased hardware lifespan by FDD, ensured persistent energy conservation performance, reduced system operational costs, conservation of historical operating records, better building operator training, reduction in required commissioning reporting in Leadership in Energy and Environmental Design (LEED) and the Energy Performance in Buildings Directive (EPBD) certification, and ensured optimal operation. Finally, tools reduce commissioning cost and improve the quality of the work by lowering commissioning effort and human error

In the case of LEB commissioning, work is often more sophisticated than with that of conventional buildings.

4.3.2 Features of EB and LEB Cx Tools Developed in Annex47

Automated and semi-automated Cx tools were developed in Annex 47. Detailed features of major tools are explained according to the structure in Section 4.3.1 and listed in Appendix 4. Table 4-3 summarizes the tools, their Cx purpose, targeted system, component, and targeted building type. FDD is the main purpose of this selection of Cx tools, followed by optimization, design and data handling. As for the Cx target, the system level is the primary target, followed by whole building and component level at the same rate, then control level. Most tools are developed for existing building (EB) rather than low energy building (LEB), as automated Cx tools for LEB Cx are currently in development.

To understand the function of a Cx tool, as well as to inform tool developers of a tool's requirements, it is important to list the type of data necessary and how it is processed. To this end, IDEFØ diagrams in Appendix 4 are useful.

The data used for each tool are further categorized based on two features: one feature is based on elements from which data are taken or related; namely, building, room environmental requirement or criteria, weather, energy, HVAC system, HVAC component, control system and maintenance. The other feature is based on the stages when data are taken; namely design stage, operational stage and measured stage. According to this categorization, data used for each tool are tabulated as shown in Appendix 5.

Table 4-3 Tool purpose, target and building type

No	Tool name	Purpose				Cx Target				Building Type	
		Design	FDD	Optimiza-tion	Data Hand-ling	Whole Building	System	Compo-nent	Control	EB	LEB
01	DABO (Diagnostic Agent for Building Optimisation), version 2007		✓	✓		✓	✓	✓	✓	✓	✓
02	CITE-AHU		✓				✓			✓	
03	Proto-type FDD tool "i-BIG" for HVAC systems		✓				✓	✓		✓	
04	Performance Analysis Tool for Heating System		✓	✓			✓			✓	
05	A tool to be used for energy analysis and fault detection on the whole building level		✓			✓				✓	
06	Optimization Tool for Air-Conditioning System Operation Considering Thermal Load Prediction Errors			✓					✓	✓	
07	An on-going commissioning tool for VRV package systems		✓					✓		✓	
08	A tool to monitor continuously oprational data of each HVAC equipment			✓					✓	✓	
09	Initial Cx tool for HVAC system in large enclosure	✓				✓					✓
10	On-going type FDD tool with pattern recognition for HVAC systems		✓				✓			✓	
11	A Tool for optimizing the operation of heating cooling plants			✓	✓		✓	✓	✓	✓	
12	A tool for optimizing the HVAC system with ground thermal storage			✓	✓		✓				✓
13	A Tool to estimate energy baselines using Simulation			✓		✓				✓	
14	Tools for Life Cycle Energy Management (LCEM) for Buildings	✓	✓	✓		✓	✓	✓		✓	
15	Energy analysis module (on Energy Plus) for VRF air-conditioning system	✓						✓			
16	CFD coupled simulation tool for annexation system of cool/heat tube and natural ventilation	✓				✓					✓
17	Proto-type FDD tool for laboratory exhaust system			✓			✓				✓
18	Energy Performance Commissioning of Existing Buildings and Building Portfolios		✓			✓				✓	

4.4 Case Studies for Various Approaches

4.4.1 German ModBen Project – Top-Down Approach

General Approach – a four-step procedure

The German ModBen project deals with the systematic performance evaluation of existing non-residential buildings. This can be regarded as an ongoing commissioning process. ModBen evaluates the developed methods and tools in six demonstration buildings.

Introducing an ongoing commissioning process to a building is a multi-level process. In ModBen the process is a top-down-approach that starts at building level and descends to selected single components, if necessary. The building's available information (stock data and measurement) will increase from step to step, allowing more detailed analysis of building performance.

A four-step procedure was developed for ModBen. Figure 4-25 illustrates the procedure on a time scale.

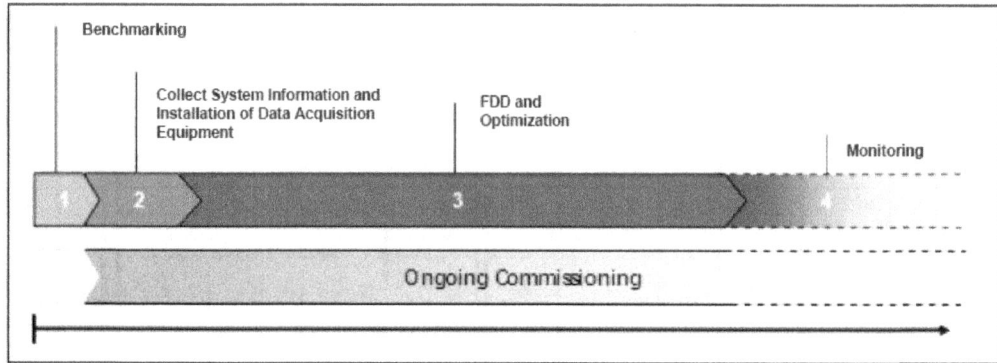

Figure 4-24 The 4-step procedure on a time scale

It is important to note that the ongoing commissioning approach may be introduced after Step 1 if a classification of the building is available at that stage.

Each step is described in detail with the definition of necessary stock data, measured data, kind of analysis, further actions, and outcomes (Table 4-4). Furthermore, tools were developed for every step (see Table 4-5).

Table -4-4 Overview of 4-step procedure

	Step 1	Step 2	Step 3	Step 4
Name	**Simple Benchmark**	**Certification**	**FDD + Optimization**	**Regular Inspection**
Description	Gather basic consumption and stock data and first classification / baseline of the building performance	Certification according to national implementation of the EPBD.	Analysis of the building performance, identification and implementation of energy saving measures and optimization of performance	Maintain optimized performance by ongoing (minimal) monitoring
Stock Data	minimal building description	According to DIN 18599	Additional stock data only according to individual needs	None
Measured Data	Utility bills / own meter readings (yearly / monthly)	None	Minimal dataset Additional measurements only according to individual needs	Minimal data set
Analysis	Benchmarks / signatures	Detailed Energy Balance (static model, monthly) Primary Energy demand Overall " COP"	Model-based Analysis ➢ Manula/automated FDD (measurement/rule-based) ➢ Manual Optimisation (model-based)	Automated outlier detection of energy consumption
Further Actions	None	Installation of data acquisition (sub-hourly data)	Implementation of energy saving measures	None
Outcomes	First classification + baseline (yearly/monthly)	➢ Certificate ➢ Theoretical target value ➢ Sub-hourly data available	➢ Faultless / optimized operation ➢ documentation of energy savings and cost	Persistence of energy efficient performance

Table -4-5 Overview of tools

	Step	ModBen-Tools
1	Benchmarking	Check list 1 (Excel based checklist for data gathering)
2	Collect system information and installation of data acquisition equipment	Check list 2 (Excel based checklist for data gathering) and EnEV+ (German implementation of EPBD)
3a	FDD (Measurement based)	Predefined Visualization (manual)
		Rule-based system (under development) (automated)
3b	Optimization (Model based)	DIN V 18599 (ennovatis EnEV+)
		Simplified simulation-model (under development) (IDA-ICE)
4	Monitoring	Automated outlier detection (regression model + clustering of day types)

Deployment of sensors

In the ModBen framework a quantified set of measured data must be delivered from every demonstration building. This set was established as it was believed to be the minimum amount of measured data necessary to facilitate a rough overall assessment of the system's performance.

The minimum data set is listed in Table 4-6.

Table 4-6 Minimum data set of measurements

Item	Measured value	Remarks
Consumption	Total fuel consumption	e.g., gas, oil, biomass
	Total consumption of district heat	
	Total consumption of district cold	
	Total consumption of electricity	
	Total water consumption	
Weather	Outdoor air temperature	From own weather station or weather data provider
	Outdoor relative humidity	From own weather station or weather data provider
	Global irradiation	From own weather station or weather data provider
Indoor Conditions	Indoor temperature	Choose one or more reference zone for that measurement
	Indoor relative humidity	Choose one or more reference zone for that measurement
System	Flow/return temperatures of main water circuits	Main heat/cold distribution. Main refers to distribution within the building and not to a primary distribution such as a district heating system.
	Control signals of pumps and fans	Main drives
	Supply and return air temperature of main AHUs	Only if supply air is thermodynamically treated
	Supply air relative humidity of main AHUs	Only if supply air is humidified/dehumidified

In the ModBen framework this minimum data set is to be achieved using a time resolution of 5 min to 10 min. Typically, sensors for energy and water consumption and for system temperatures are installed in the building. Sensors for weather data and indoor climate are typically only installed if the building is equipped with an air conditioning system.

However, even when sensors are installed, the transfer function to other analysis systems is generally not provided. Availability of measured data is enhanced in buildings with a BAS that comprises a management level. In theory, the BAS can provide a lot of data. Unfortunately, the BAS in most existing buildings is not equipped with data transfer features such as a data base interface, an Open Process Control (OPC) server or the ability to export ASCII files with measured values. As well, energy data is typically not available from the BAS. The technology might be available but, for cost reasons, is often not installed.

Nevertheless, if a BAS is installed it should be checked to see whether it can be modified to provide the minimal data set at a reasonable cost. If the enhancement of the BAS is complex, it might be more cost efficient to install an extra data logger with additional sensors. If no BAS is available, the installation of a data logger is the only option unless the building owner decides to upgrade the building with a BAS.

Another aspect to observe in high time data logging (hourly or sub hourly) is the amount of traffic the fieldbus must handle. In the case of significant amounts (>> 100 data points) modifications to the fieldbus communication are necessary.

In most cases, heat and water meters must be equipped with a pulse output or fieldbus interface. In the case of a fieldbus interface, they must also be equipped with a grid-connected power supply because of frequent readings.

In the ModBen project, the cost of acquiring the minimal data set is 15 000 EUR to 60 000 EUR per building, depending on the system's actual condition. However, in most cases the cost averages a reasonable 5 % to 20 % of annual energy costs.

The effort to acquire additional data that exceeds the minimal data set relies on the presence of a BAS. If a BAS is enhanced to deliver the minimal data set, additional data from the BAS (e.g., control signals, schedules, etc.) is available at little cost but if a data logger is used, acquiring additional data results in significantly greater cost.

Table 4-7 gives an overview of the cost of the data acquisition system in the ModBen project demonstration buildings. From our experience, we can conclude that:

- If the main part of the measurement equipment and a BAS are installed, the BAS should be enhanced for monitoring, especially if many datapoints (>>100) are to be recorded.
- If only a few measurement points are required and/or no BAS is installed, a separate data logger is the better solution.

Table 4-7 Overview of data acquisition cost and estimated static amortization (red = estimated; blue = calculated ; ??= unknown)

		DVZ Barnim	Großpösna	M+W Zander Zentrale	EADS 88	LEH	Kraft Foods
BAS and measurement equipment already available	-	Yes	No	Yes	Partially	Yes	Partially
Kind of data acquisition	-	BAS	Data Logger	BAS	Data Logger	BAS	Data Logger
Number of data points		300	20	500	80	480	450
Annual energy cost	EUR	63 000	4 900	617 000	150 000	360 000	??
Cost of data acquisition and gathering system	EUR	30 000	3 000	21 000	30 000	15 000	60 000
Cost of data acquisition and gathering system	% of annual energy cost	48 %	61 %	3 %	20 %	4 %	??
Expected minimal energy savings	%	5 %	5 %	5 %	5 %	5 %	5 %
	EUR/year	3 150	245	30 850	7 500	18 000	??
Static pay back	years	9.5	12.2	0.7	4.0	0.8	??

Performance metrics

ModBen applies and evaluates different pre-defined performance metrics according to the top-down approach described. The next sections give a short general description while the next chapter provides examples from demonstration buildings.

- Statistic Benchmarking (annual/monthly data in Step 1)

  Several sources exist for reference values for energy consumption of non-residential buildings in Germany. The values are statistically derived from building stock. Typically, they are listed according to the building's main use (office, school, shop, etc.). Unfortunately, most of these benchmarks miss detailed specifications, such as year of construction, building size or HVAC system type, which reduces the usefulness of these benchmarks.

- Building specific target value

  According to the EPBD, certification delivers a building's specific target value for energy demand. German implementation of the EPBD (DIN V 18599) is complex and includes most HVAC systems. The building is further divided into multiple zones according to use.

Although certification works with standard use profiles, the calculation model can be adjusted to the building's actual use (operation and occupancy schedules, setpoints) – to a certain extent. This "calibration" requires considerable effort and results are not directly comparable to measured energy consumption

- Pre-defined visualization of minimal data set for manual fault detection and diagnosis
  Standard visualizations for the minimal data set were defined in order to extract information from the data. Visualization includes carpet plots for checking time patterns of operation and occupancy and scatter plots for identifying control strategies and time series as reference and for detailed examinations.

  These plots can be manually inspected in order to find faults and deficiencies when analyzing building performance for the first time.

- Rule based fault detection and diagnosis
  The variables of the minimal data set will show different correlations for faultless operation and operation with faults. Theoretically, it is possible to calculate the correlations for each case to be able to identify faults from real measured data.

  ModBen is still developing a rule based system which is able to detect these typical faults from measured data but this function will not be discussed further in this report.

- Simplified simulation models for optimization
  Numerical Optimization requires simulation models that run fast and so enable many variations. ModBen develops simplified models to reduce the numerical effort required for simulation. These models can be calibrated with real measured data and used for optimizing the control strategy of real buildings. This feature remains under development.

- Automated outlier detection for ongoing monitoring and maintenance of fault free/optimized operation
  If acceptable or even optimal operation of an HVAC system is achieved, it will still require maintenance and monitoring. Building operation or performance must be monitored weekly or daily. As operational staffs do not have time for this, ModBen has developed a method to automate the process.

  The building's current performance is identified first using a regression model, which is then combined with a clustering process that identifies day types and consumption patterns.

The calibrated model is then used for a daily automated check of energy and water consumption.

Building case study

Examples show how measured values of the minimal data set are used to analyze or monitor the building. Not all analysis routines or performance metrics are presented, just the visualization (manual fault detection) and automated outlier detection.The example used for data visualization is the ThyssenKrupp Kreuzgebäude in Essen shown in Figure 4-26.

Figure 4-25 ThyssenKrupp Kreuzgebäude in Essen, Germany

The Kreuzgebäude is a typical office building built in the 1980s with an insulation standard (mean U-value= 1.3 W/m²K, percent of façade's glazing = 30 %). The building's energy supply is provided by district heat and electricity.

The building is heated with warm water radiators; most offices are ventilated by openable windows. Roughly 25 % of the spaces are equipped with a central ventilation system with heat recovery and a heating coil for preheating of air.

Approximately 30 % of the spaces are equipped with split units for cooling in summer.

The building's annual energy cost is about 150 000 EUR.

Data acquisition

With Kreuzgebäude no BAS was available, so the minimal dataset is acquired using a data logger as depicted in Figure 4-27. Necessary sensors were installed with most communication done via the modbus protocol. To acquire 85 data points costs roughly 25,000 EUR.

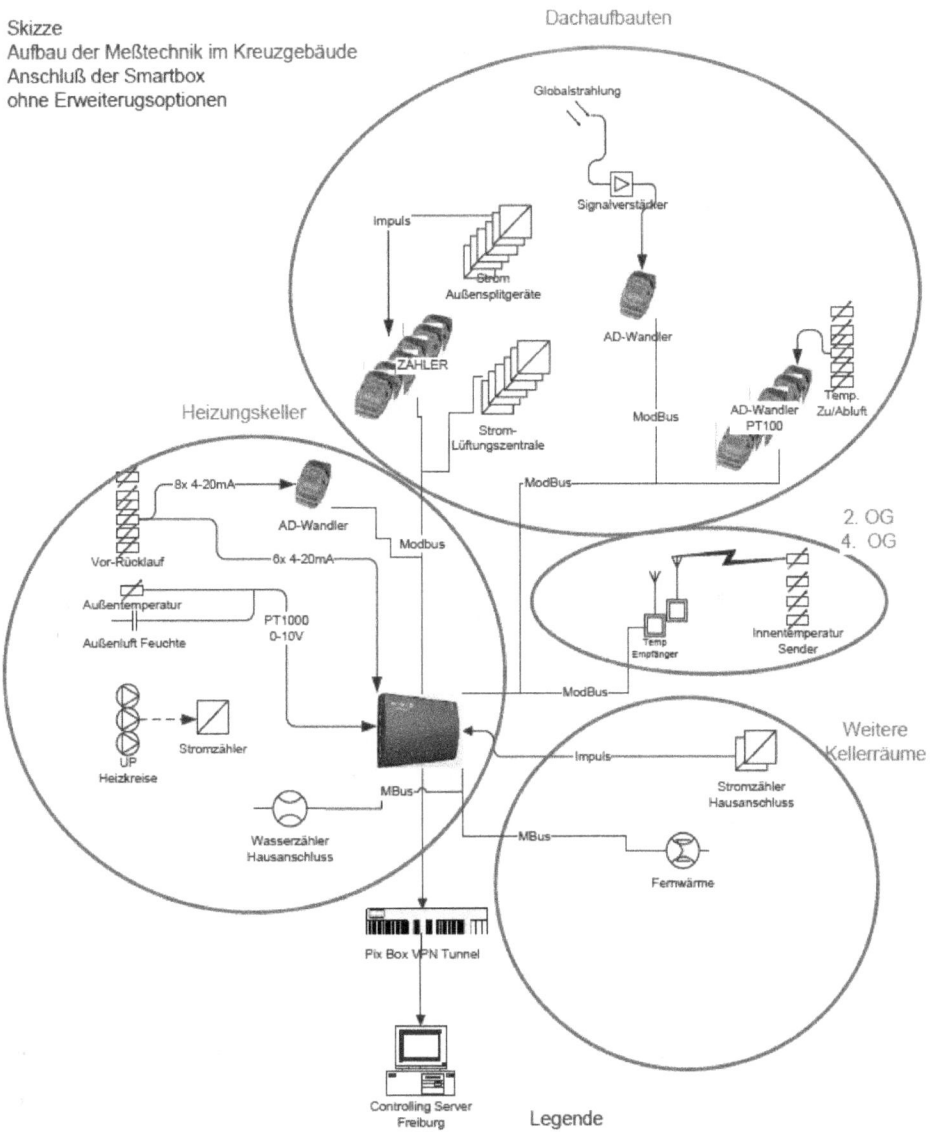

Figure 4-26   Data acquisition system installed in Kreuzgebäude. Blue represents the data logger. Data is transferred to Fraunhofer ISE via internet (VPN tunnel).

Visualization

Manual fault detection using pre-defined data visualization is done in a top-down manner, from overall energy consumption to system/subsystem levels.

The plot in Figure 4-28 illustrate some findings in Kreuzgebäude.

Energy consumption

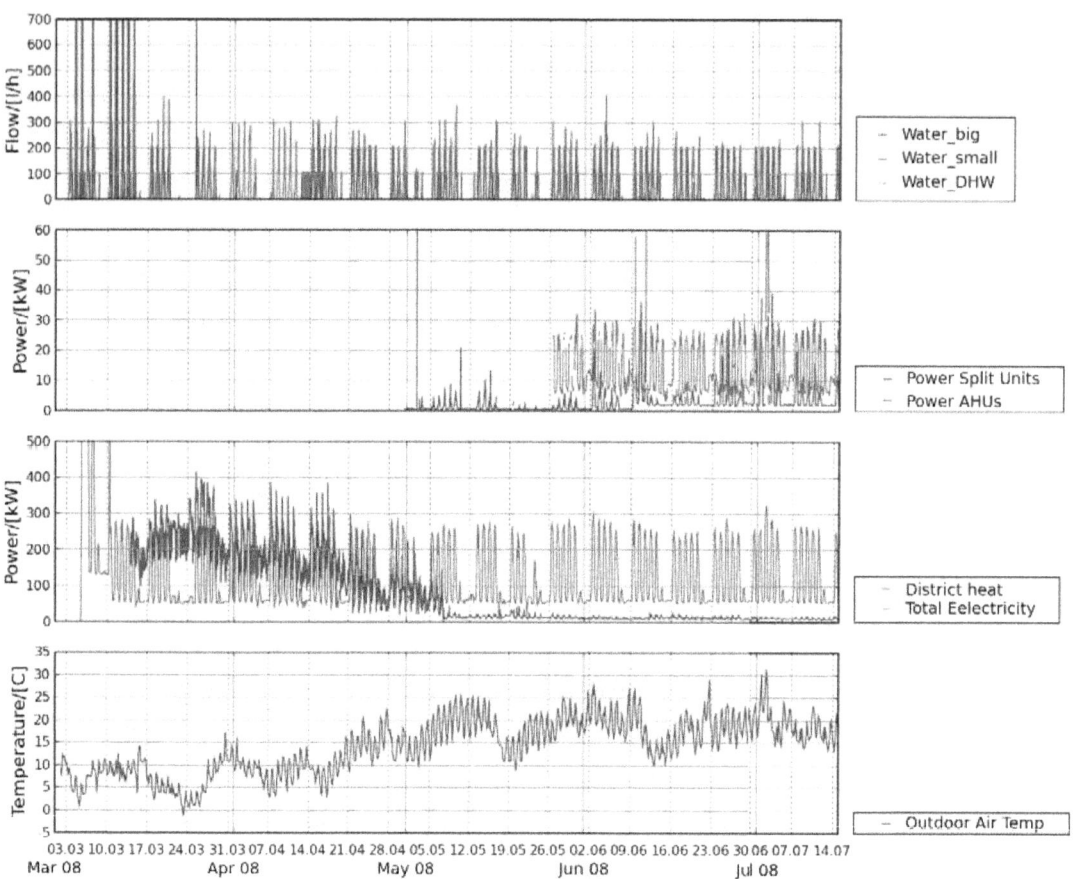

Figure 4-27   Time series plot of energy and water load (hourly values)

Figure 4-28 shows the time series plot for energy and water consumption together with outdoor air temperature. Although weekly patterns and the dependency of heating consumption on outdoor air temperature is plotted, the following figures show that additional information can be extracted. Time series plots are used only as reference plots.

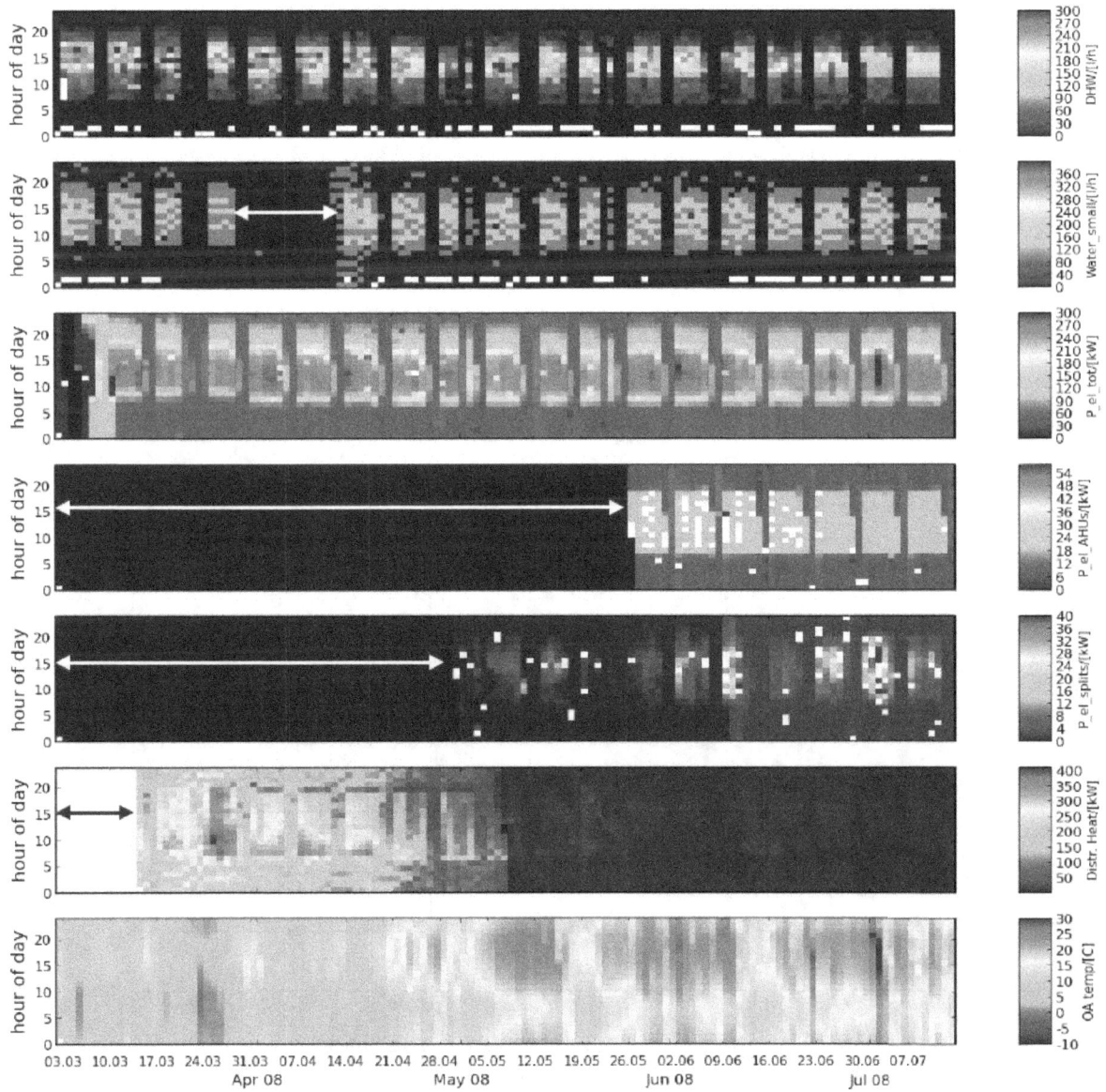

Figure 4-28 Carpet plot of energy and water load (hourly values) from bottom to top: outdoor air temperature, district heat, electricity split units*, electricity ventilation*, total electricity, total water consumption, DHW consumption. (*not part of minimal data set), arrows show missing data

Figure 4-29 shows carpet plots of energy and water consumption. Daily and weekly patterns are more obvious than in Figure 4-28. The building's occupancy – from water and electricity consumption – is Monday to Friday, 07:00 to 19:00. The electricity plot shows energy consumption Saturday mornings as well, which seems unnecessary as water consumption (and occupancy) is zero at that time. The district heat plot reveals that the night and weekend setback is deficient, at least at low outdoor air temperatures.

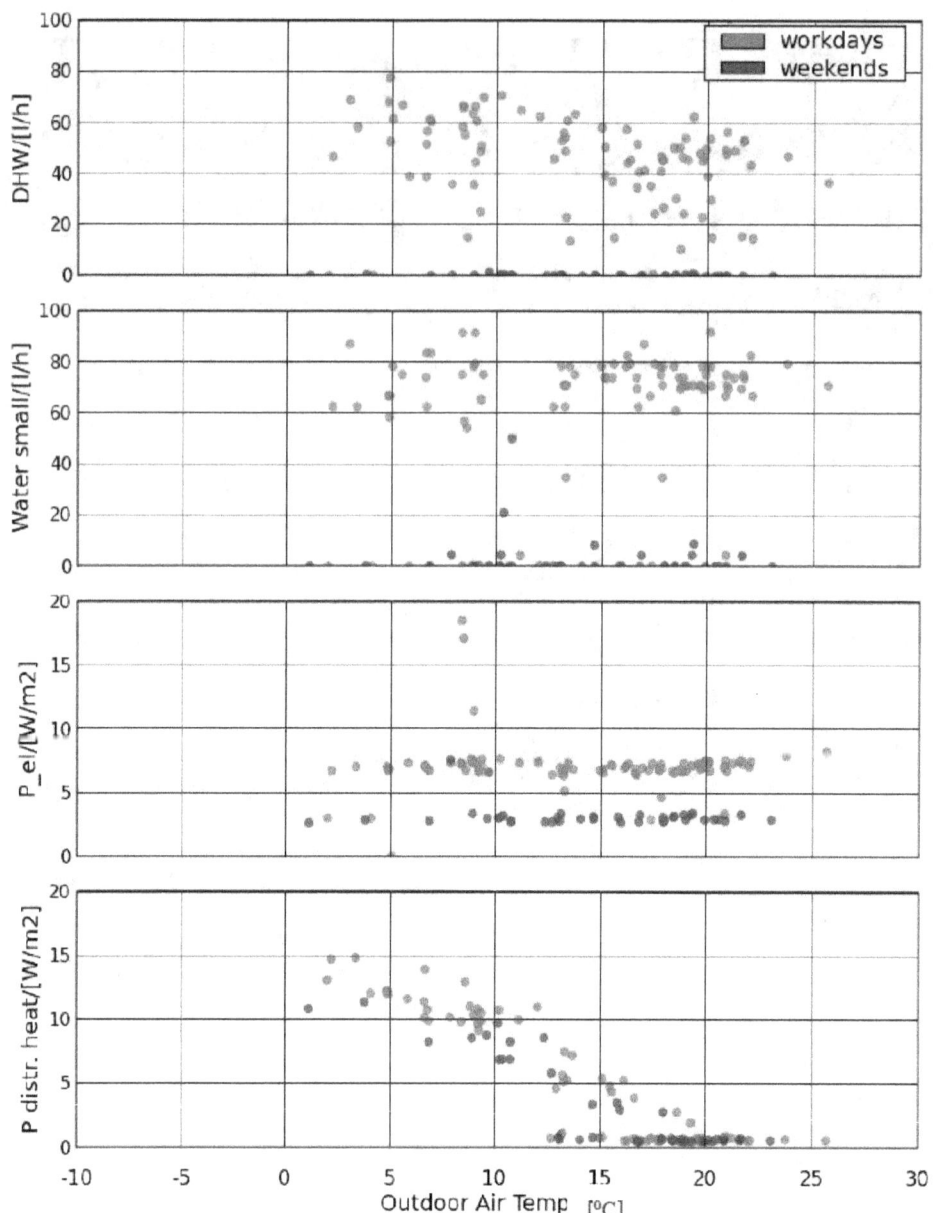

Figure 4-29   Energy and water load signatures (daily means) by weekdays and weekends.

Figure 4-30 is a scatter plot (signatures) of energy and water consumption in relationship to outdoor air temperature. The missing set back of heating during weekends is obvious in the lowest plot. This contradicts electricity and water results that show low or no occupancy on weekends.

The electricity plot shows that cooling is no major consumer as the electricity load increases only slightly with higher outdoor air temperatures.

Heating circuits

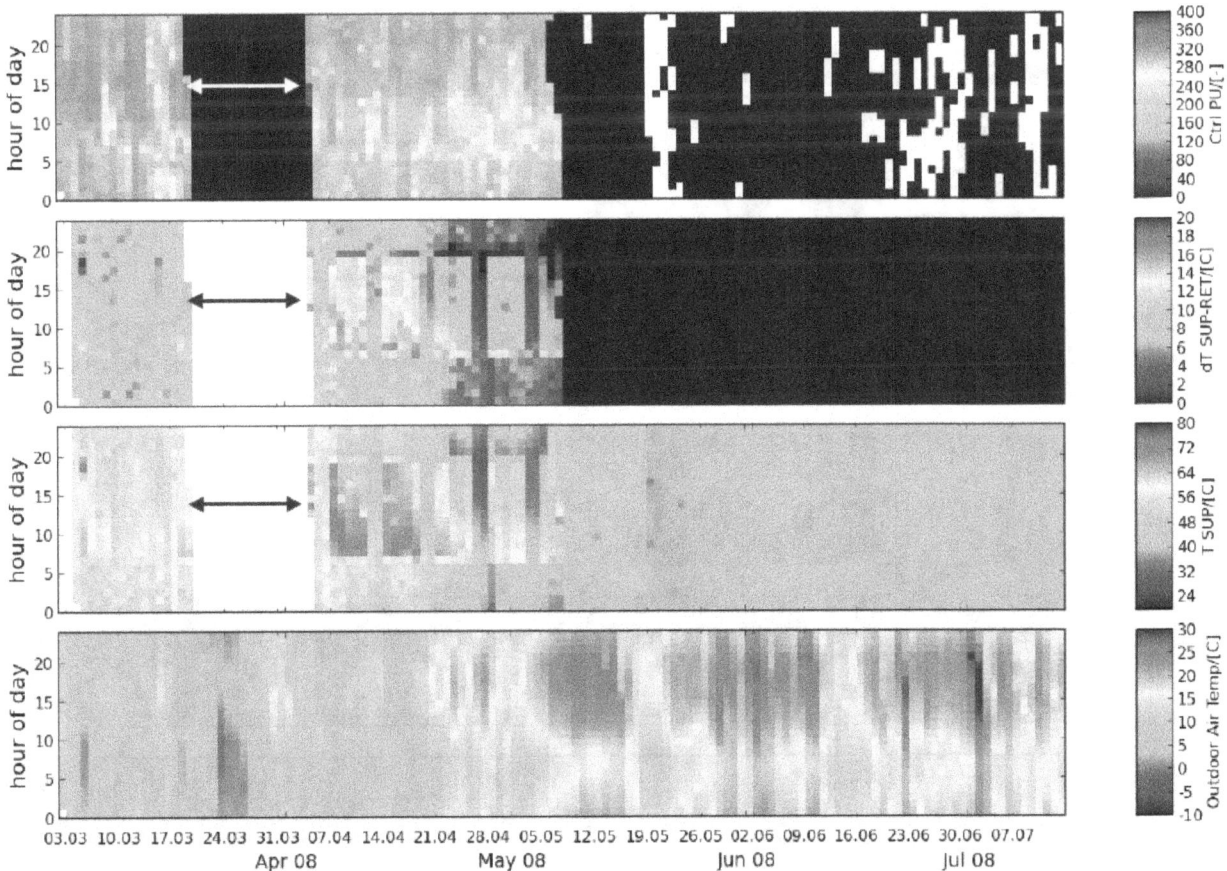

Figure 4-30   Carpet plot of heating circuit sensors (arrows show missing data)

Figure 4-31 shows a carpet plot for one of the building's four heating circuits. Arrows overlaid on the carpet plot for the supply temperature and temperature difference show the same deficiency in night and weekend setback that was already recognized for the heating load in Figure 4-29.

Also, the pump is in constant operation, representing another potential for optimization. The pump could be turned off during unoccupied periods and changed to variable flow.

The plot also shows that the heating is turned off (in May) regardless of the outdoor air temperature.

Air Handling Units

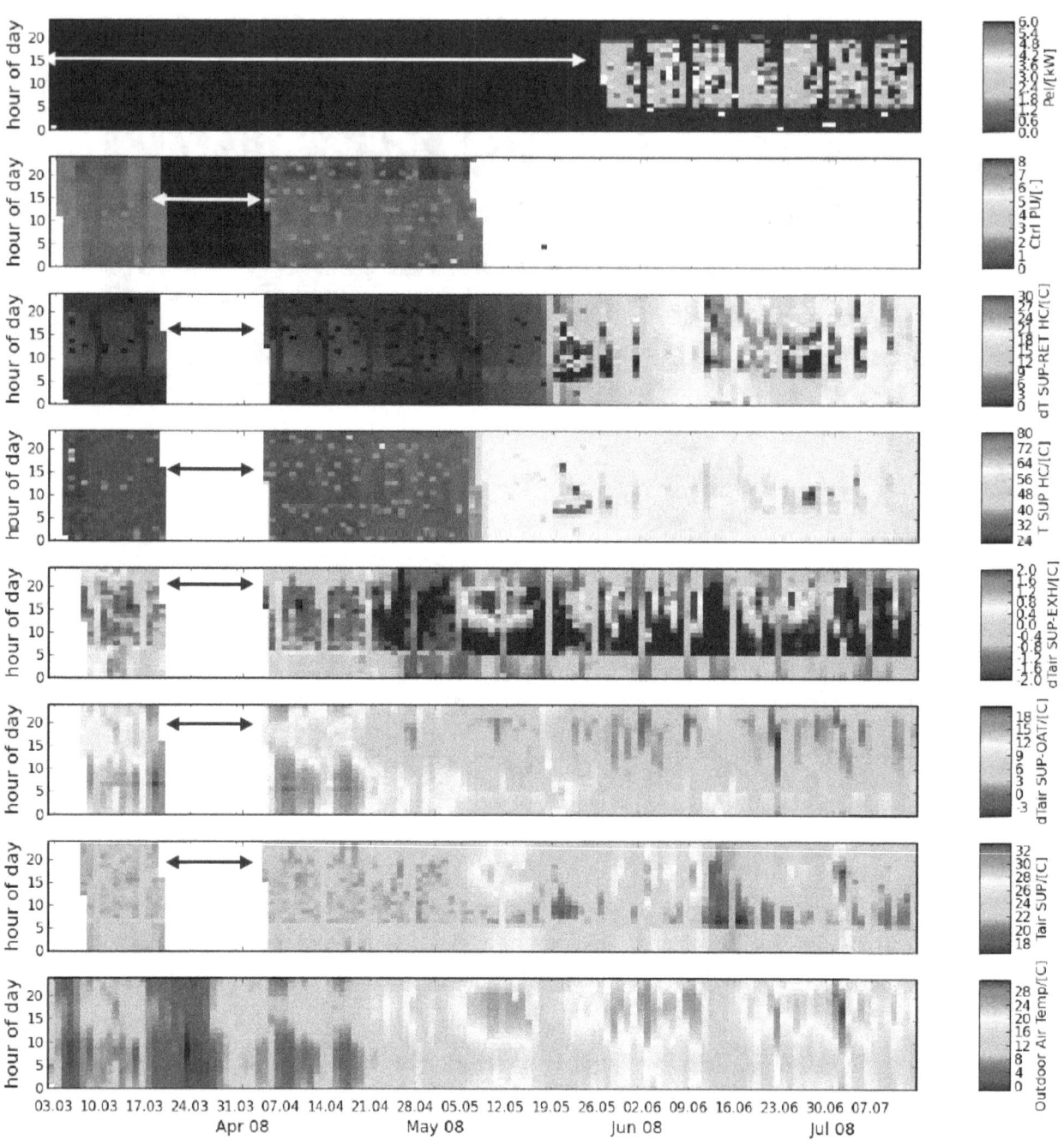

Figure 4-31   Carpet plot of air handling unit sensors arrows show missing data

Figure 4-32 shows a carpet plot for one of the air handling units. The operation schedule is Monday to Friday, 06:00 – 21:00 and Saturday 06:00 – 14:00. (Electricity – the top plot - is used here as the AHU operational signal)

78

Saturday operation is unnecessary and will be turned off.

The heating coil's primary pump is in constant operation during the entire heating season and is turned off manually in May. Like the pumps for the heating circuits the pump could be turned off during unoccupied periods and changed to variable flow.

The temperature difference between the heating coil's supply and return temperature (dT SUP-RET HC) shows odd fluctuations in summer – even though the primary pump is turned off. This occurs because the secondary pump (see Figure 4-33) starts up as soon as the outdoor air temperature falls below 20 °C to avoid freezing. This causes unnecessary heat loss and heats up the air supply in summer, which causes extra cooling load. This operation has to be changed. It will be tested to turn off the primary pumps completely as the secondary pumps are obviously capable of supplying the heating coil. This fault impacts building optimisation. Figure 4-34 shows that the temperature difference on the water side of the heating coil is very low (2 °C to 3 °C). This means that the primary pump is set to over capacity.

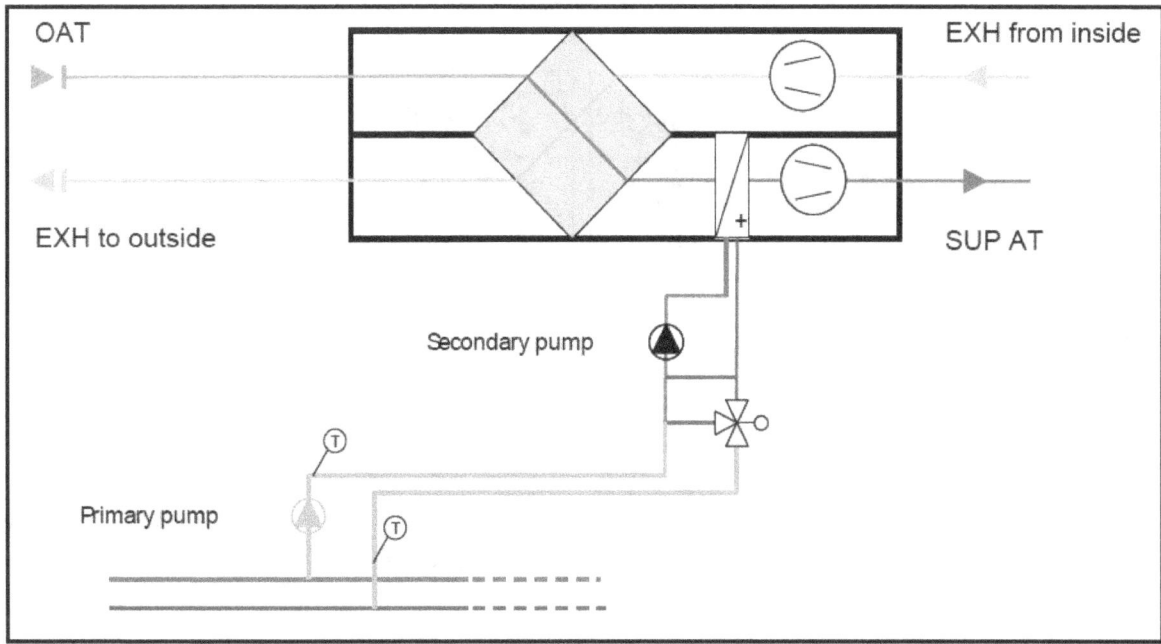

Figure 4-32   Layout of air handling units

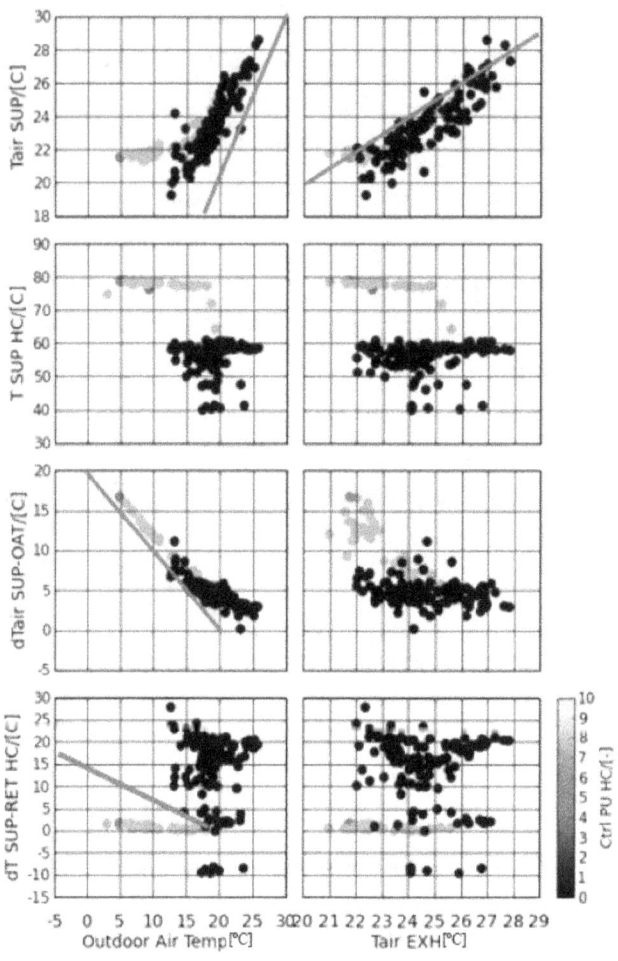

Figure 4-33   Scatter plot air handling unit sensors. Color scale represents
the control signal of the heating coil pump.

Results and Conclusions

In the best-case scenario, manual fault detection by sensor visualization in the minimal data set revealed energy savings of up to 15 % of annual energy consumption.

Using a typical minimal data and frequent faults (basic faults make it easier to detect) the following were identified:

- Insufficient or no setback of heating and cooling during nights, weekends or unoccupied periods;
- Missing adjustment to operating schedules between sub-systems in the HVAC system;
- Heating/Cooling curve incorrectly set;
- Mass flow rates too high/over capacity;
- Wrong size of generators; and
- Faults in measuring equipment;

Automated fault detection

80

Once an optimal or acceptable performance is reached, it must be monitored to maintain it.

As this monitoring would be tedious for operational staff a method was developed to automatically detect abnormal energy consumption values on a daily basis by using the minimal data set.

The method is based on a regression model combined with a clustering process for identification of day types, i.e., days with varying consumption profiles in all delivered energy and water. In Germany this will typically be heat (e.g., gas or district heat) and electricity.

The regression model is trained using data that contains at least three months of measurements. The training identifies basic patterns in energy and water consumption. After training, the model can be used for daily fault detection (validation phase, see Figure 4-35).

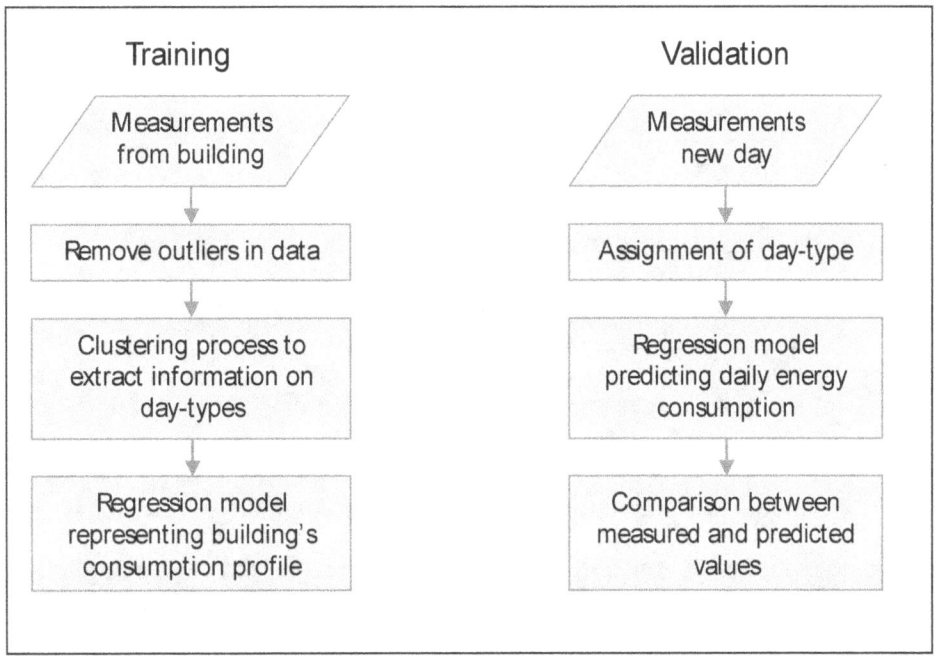

Figure 4-34   **Automated fault detection method with training and validation phases**

With two test buildings it was shown that the method can identify the consumption profile reasonably well and detect real and manually inserted outliers. The method will be further tested on other buildings.

Figure 4-36 shows a comparison of the measured values of heat load and the prediction of the model for one of the demonstration buildings.

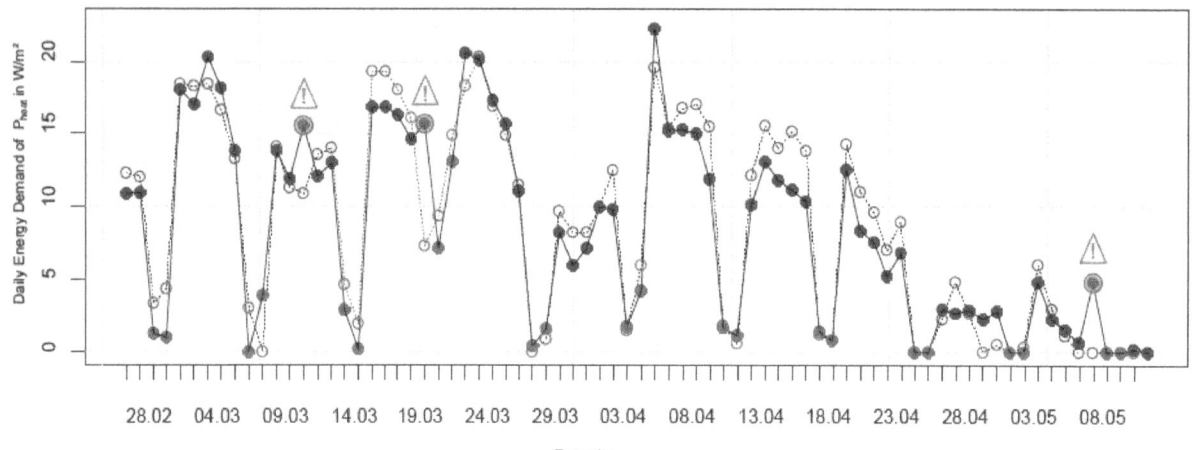

Figure 4-35 **Comparison of daily heating demand at demonstration building with estimated values of regression model**

Details of the black box model are as follows:

- Daily heating demand is modeled by applying a black box model which can also be developed to monitor electricity or water demand

- Black box model contains clustering process to automatically identify building's day-types (days with different consumption patterns)

- A multiple linear regression model estimates model parameters

- The model's estimated values are compared to the measured values at the building site and outliers can be identified on a day-to-day basis

- The model is not dependent on the building layout and can be applied without extensive prior knowledge of the construction characteristics

- Measured data from the building site is the only key component

4.4.2   Netherlands – Bottom-Up Approach

General approach

TNO Built Environment and Geosciences used a three-step approach for performance analysis of buildings and HVAC systems. This procedure follows a general bottom-up approach, concentrating on the technical and functional condition monitoring, concerning operation and maintenance of HVAC systems.

The three-step procedure

The three main steps are:

1. Simple benchmark on comfort;

2. Analysis and optimisation; and

3. Regular inspection and check of action plan.

Starting points of this approach:

- Take available (measurements) from BAS and try to use it for recommissioning; and

- Special comfort problems need to be solved (additional sensors may be needed).

The commissioning projects executed until now in 90 buildings were based for 95 % on re-commissioning projects. The main reason was to solve problems of the tenants in the buildings, which have resulted already in numerous comfort complaints.

The client approach for a performance test is illustrated in Figure 4-36.

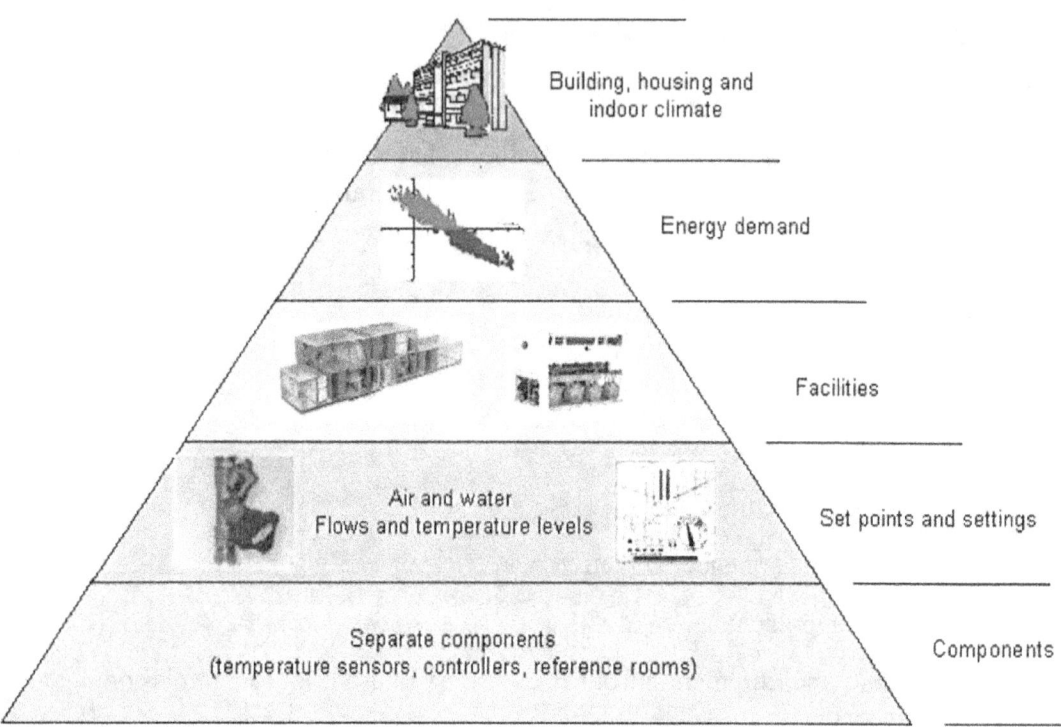

Figure 4-36   **Principle of evaluating a building's HVAC systems**

The building and the building user with its specific business processes determine the inside climate/comfort and energy demand. Whether the energy demand of the building can be met, depends on the HVAC concept and on the installation's operational management. For our investigation it was assumed that the original design was sufficient for the desired comfort in the building, so there is no preceding check of the design and capacity of the installation's components.

Some random checks of the installation and separate equipment, as shown in the lowest two tiers in the pyramid of Figure 4-36, were executed. Experience shows that when the quality of the lowest two tiers is insufficient, the upper tiers/levels are also of insufficient quality.

A TNO-study of more then 90 buildings of different types showed comfort complaints and high energy consumption to be the most common faults.

The problems are:
- The average energy use of buildings is 25 % higher than expected;
- 70 % of HVAC systems do not work as expected; and
- 90 % of complaints are attributed to poor HVAC system performance.

The causes are:
- 15 % design faults; and
- 85 % (i) completion (handover) of HVAC system and (ii) operation and maintenance.

We will focus on technical and functional monitoring aspects concerning operation and maintenance of HVAC systems.

The details of the steps

Table 4-8 presents a comfort-oriented approach.

Table -4-8 Overview of the 3-step procedure

	Step 1	Step 2	Step 3
Name	Benchmark on comfort	FDD + optimization	Regular inspection and check of action plan
Description	Compile a basic survey list of complaints and stock data and first classification / baseline of the building performance	Analysis of the building performance, identification and optimisation of performance	Check that targets identified in step 2 are achieved
Stock Data	Minimal building description	Additional stock data according to individual needs	None
Measured Data	Make a first analysis of BAS dataset. Manual checks concerning sensors and visual inspections.	Minimal dataset + additional measurements according to individual needs.	None
Analyse	Benchmark based on survey of complaints	Standard analysis System specific analysis	Step 2 targets as major metric
Further actions	None	Installation of data acquisition (hourly data)	None
Outcomes	First classification of the type of complaints	Faultless / optimized operation Documentation of improvements actions/ proposals	An improved or adjusted action plan
Tools	"Checklist 1"	- "Analysis framework"	

The Quick scan

The following section describes every step in detail. The activities executed for each building is based on a so called quick scan [39] which describes all the necessary activities, including a survey list to be filled out. The four main activities executed for each building are:

Preparation and Start-up meeting (Step 1)

A start-up meeting with the building owner will be held to discuss basic assumptions and exchange of information. We explain clearly that we concentrate on the technical and functional condition monitoring concerning operation and maintenance of HVAC systems.

Inspection of buildings and installations (Step 2)

Global inspection of buildings and installations will focus on:

1) Evaluation of design of HVAC system

   An inventory will be made of the HVAC system. The focal point of this inventory:

   - Does the original installation concept fit with the current building user and its operational management?

   - Is zoning necessary? If so, is this realised in the installation?

   - Is the outdoor air temperature sensor placed correctly, in a representative location?

   - What comfort level can we expect with this installation concept?

   - Can we expect with this building design some limitations in the expected (realised) comfort?

2) Inventory of the realised comfort conditions

   Manual checks will be executed concerning:

   - Room air temperatures;

   - Room radiation temperatures; and

   - Air velocity.

   The question to be answered is: can we expect these measured values when the installation is working optimally?

3) Design use versus actual use

Design building use:

- Which internal heating load was taken into account?

- Type of organisation / operational management?

- Which type of room functions were defined and assigned?

Actual building use:

- Does the current function of the rooms / zones correspond the original design?

- Is it possible to restructure the building interior without changes to the installation?

- If sensors are installed, are they placed in a representative place and position?

4) Performance of the HVAC system

The performance of the HVAC system will be checked on three levels:

- Production of heating and cooling (boilers, chillers)

- Distribution of heat and cold in the building (water and air)

- Emission of heat and cold in the zones and rooms.

The type of checks executed on the three levels:

- Is test, adjust and balancing (TAB) executed in conformance with the design?

- Check of temperatures (water and air)

- Reliability / calibration of temperature sensors in the zones

- Performance and adjustment of controllers

Note:We use as much as possible, including data point information collected by the BAS/ BEMS and available monitoring (TAB) reports.

Analysis and reporting (Step 2)

The following analysis and reporting items will be executed:

1. Analysis of the inspections

2. Short reports of analysis and conclusions

3. Causes of comfort complaints

4. Suggestions for recommissioning of building and HVAC system/ installations

Final meeting and discussion (Step 3)

A final meeting per building is planned to explain and clarify the results and the actions to be taken.

Building Case Study 1

Benchmark on comfort

A simple benchmark on comfort is presented in Figure 4-38. The key performance indicator, complaints per person per year is calculated as follows: in Building 1 are 1500 persons and 1080 complaints were received ovet the year. Therefore, 1080 complaints/1500 persons = 0.72 complaints per people per year. From 32 buildings, an average key performance indicator is derived of 16 % comfort complaints per person per year. After resolution, a target value of 8 % is realistic.

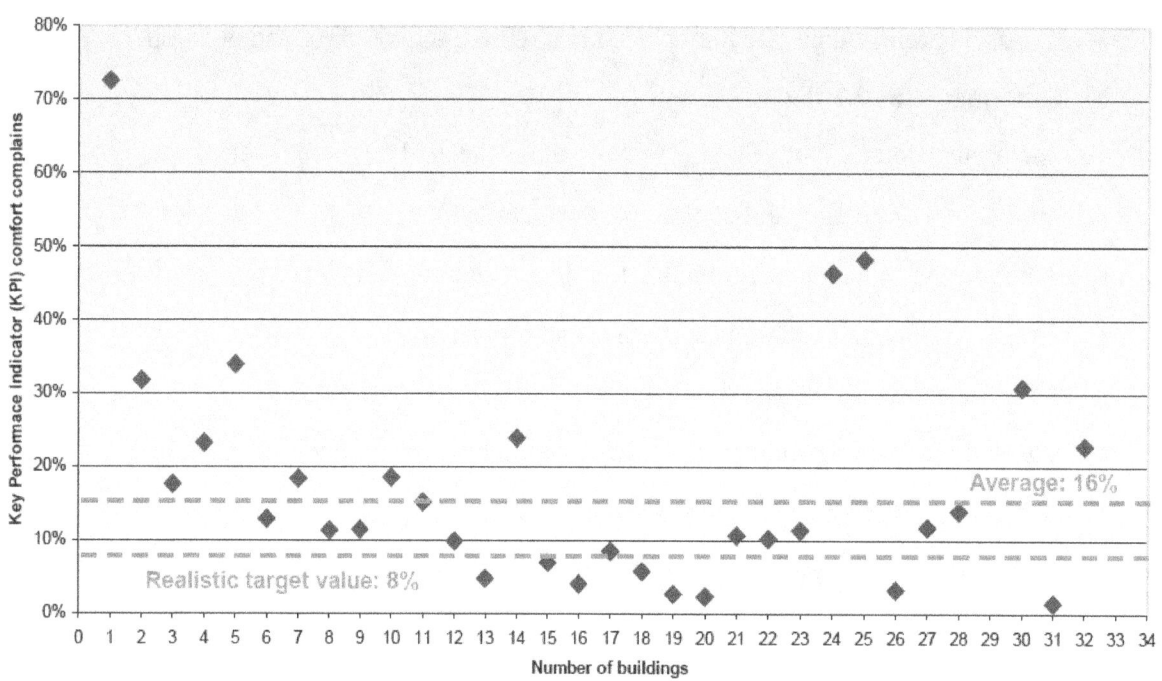

Figure 4-37   A simple benchmark on comfort based on survey lists of complaints

- Are comfort improvements realistic?

Our task is successful if the complaint list of the building user is strongly reduced.

An example is given in Figure 4-38.

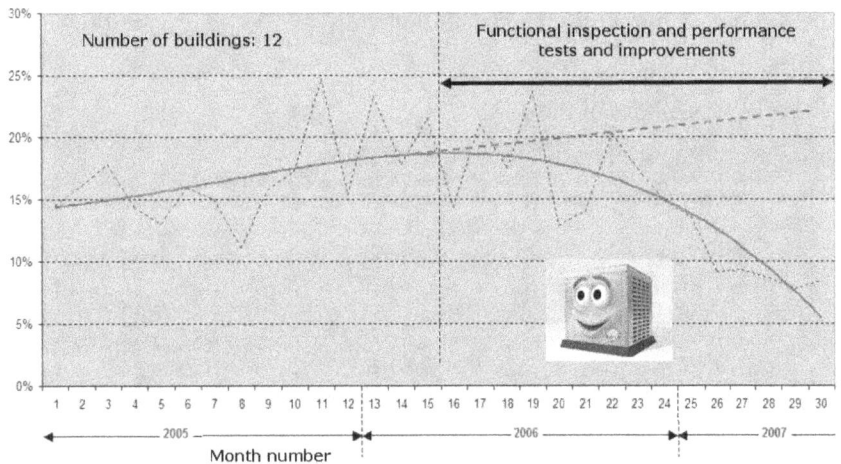

Figure 4-38   Reduction of user complaints after improvements executed by TNO.

In this example the complaints are from 12 building of a insurance company. The benchmark on comfort showed comfort-related complaints were at a level of 20 % in those 12 buildings. After improvements, the measured reduction was between 5 % and 10 %.

Building Case Study 2

Deployment of sensors

A building owner in the Netherlands gave permission to investigate the real performance of one of the AHU's in his building. It was executed free of charge for the building owner in a so-called SME project. One objective was to convince installers and service companies of the benefits of FDD of HVAC systems.

The building selected was a hospital in the town of Almere (Figure 4-39). The AHU was selected by the building owner.

Figure 4-39   Plan of the hospital in the town of Almere

The starting point of the performance test is a blind test, based on:

- Functional description (6 pages) of HVAC system and selected AHU;
- Principal layout of the AHU (Figure 4-42);
- No additional sensors installed;
- Use of available sensors in the BAS;
- Data collection prepared by building owner; and
- Excel files forwarded by e-mail to TNO.

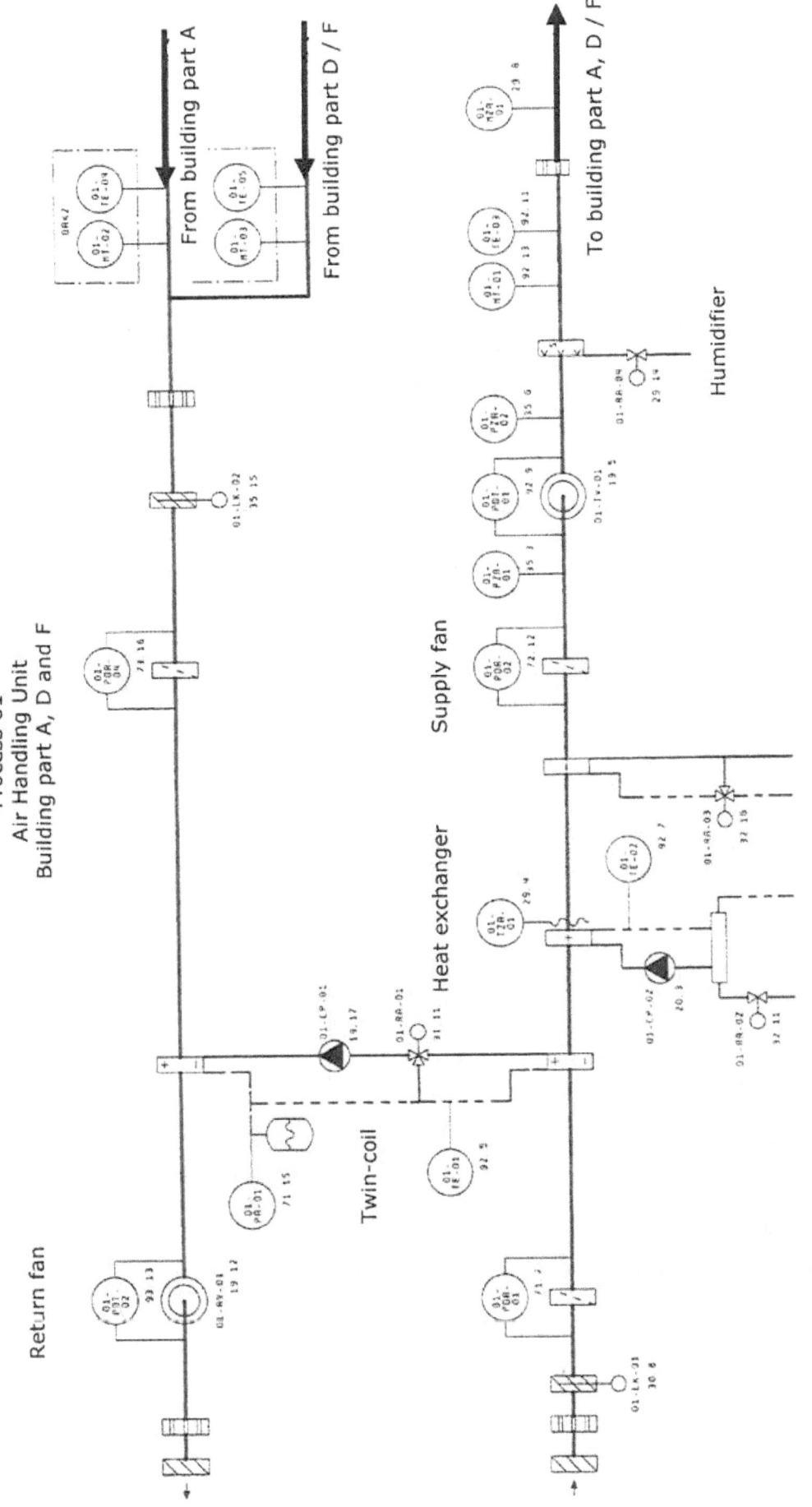

Process 01
Air Handling Unit
Building part A, D and F

From building part A

From building part D / F

To building part A, D / F

Return fan

Supply fan

Heat exchanger

Humidifier

Twin-coil

Figure 4-40   The layout of the AHU.

91

Table 4-9 presents an overview of the points measured in the AHU.

Table 4-9 Data set of measurements

Flevo Hospital, city of Almere, The Netherlands
Date : 8 March 2007
Subject : AHU measuring points
Data collection : every 5 minutes for benefit of FDD test

Item	Code in drawing	Measured value	Unit	Remarks
Weather		Outdoor air temperature	°C	own weather station
		Outdoor relative humidity	%	own weather station
Indoor conditions	01-MT-02	Indoor air temperature	°C	Use return air temperature building part A
	01-TE-04	Indoor air relative humidity	%	Use return air rel. humidity building part A
	01-MT-03	Indoor air temperature	°C	Use return air temp. building part D+F
	01-TE-05	Indoor air relative humidity	%	Use return air RH building part D+F
Generation	Chillers	Chilled water temperature Chiller 1	°C	Central cold water production
		Chilled water temperature Chiller 1	°C	Central cold water production
	Heating	Supply water temperature of main central heating system	°C	Central hot water production by gas boilers
AHU		Set-point absolute humidity, supply air.	g/kg	
		Set-point absolute humidity, return air.	g/kg	
	01-CP-01	On-Off circulation pump twin-coil	1/0	
	01-CP-02	On-Off circulation pump heat exchanger	1/0	
	01-AV-01	Operation return fan 100 % or 50 %	1/0	
	01-TV-01	Operation supply fan 100 % or 50 %	1/0	
	01-TZA-01	Frost protection thermostat	1/0	
	01-MT-01	Relative Humidity supply air	% RH	
		Lowest measured outside air temperature of all outside sensors	°C	
		Temperature set-point supply air	°C	
	01-TE-03	Temperature supply air	°C	
	01-TE-02	Return water temperature of heat exchanger	°C	
	01-TE-01	Supply water temperature twin-coil	°C	
	01-RA-03	Control device (valve) cooling coil	%	Control signal
	01-RA-04	Control device (valve) humidifier	%	Control signal
	01-RA-02	Control device (valve) heat exchanger	%	Control signal
	01-RA-01	Control device (valve) twin-coil	%	Control signal
	01-PT-01	Pressure drop return fan	Pa	
	01-PT-02	Pressure drop supply fan	Pa	

The operation matrix of the AHU is presented in Table 4-10.

Table 4-10 The operation matrix of the AHU

AHU building part A, D and F	Operation and Control AHU							
AHU components	1	2	3	4	5	6	7	8
Fresh air damper	Fc	Fo	Fo		Fo	Fc		
Circulation pump twin-coil	Aut	Aut	Aut		Aut	Aut	On	
Control device (valve) twin-coil	0	Co	Co		Co	0		
Circulation pump heat exchanger	Aut	Aut	Aut		Aut	On	On	On
Control device (valve) heat exchanger	0	Co	Co		Co	100	Co	Co
Control device (valve) cooling coil	0	Co	Co		Co	0		
Supply ventilator LOW (50 %)	Off	Off	On		Off	Off		
Supply ventilator HIGH (100 %)	Off	On	Off		On	Off		
Steam control device (valve)	0	Co	Co		Co	0		
Return ventilator LOW (50 %)	Off	Off	On		Off	Off		
Return ventilator HIGH (100 %)	Off	On	Off		On	Off		
Exhaust air damper	Fc	Fo	Fo		Fo	Fc		
Control								
Supply air control	Of	Co		Co		Co	Off	
Supply air humidity control	Of	Co		Co		Co	Off	
Return water temp. control heat exchanger	Of						Co	Co

Explanation operation management		Operation situation		Exceptional operation situat.	
Off	Switch Off	1	OFF	5	Fire
On	Switch On	2	IN Daily operation	6	Frost protection
Fc	Fully closed by controller	3	IN Night operation	7	T-Outside < frost limit
Fo	Fully open by controller	4		8	Start up AHU
Co	Regulate/ control by DDC				
Aut.	Off / On through DDC				
Aut.	Closed / Open through DDC				
L	Low				
H	High				

The AHU set-points and operation schedule are presented in Table 4-11.

Table 4-11 AHU set points and operation schedule

Set up points	Set-points
Desired supply air temperature when return air temperature < 22 °C	20 °C
Desired supply air temperature when return air temperature > 26 °C	16 °C
Desired return temperature heat exchanger during start up situation	30 °C
Frost protection when outside air temperature is lower then:	5 °C
Frost protection twin-coil	8 °C
Enabling cooling when outside air temperature above:	15 °C
Xp control twin-coil	10
Xp control heat exchanger	10
Xp control cooling coil	10
Tn temperature control	3 min
Enabling steam humidifier after AHU start after:	10 min
Enabling steam humidifier when outside air temperature below:	15 °C
Desired return air humidity	6.2 gr/kg dry air
Xp control steam humidifier	10
Xn control steam humidifier	3 min
Maximum control signal steam valve	50 %
Maximum humidity content	9.0 g/kg dry air
Minimum pressure drop for testing belt break (after 2 minutes)	20 Pa
Clock settings	
Operation period AHU in LOW speed (50 %)	Continue
Operation period AHU in HIGH speed (100 %)	Monday till Friday from 08:00 till 22:00

Performance metrics

- Definition of expert rules

The analysis approach is based on fault detection and fault diagnosis. Using information from the AHU's principal layout (Figure 4-40) and Table 4-9 to Table 4-11, expert rules are derived. Table 4-12 presents an example of the algorithm of six knowledge rules.

Table 4-12 Algorithm knowledge rules

Fault	Knowledge rule	Description
1	IF mode=1: $\|T_{sa} - T_{sa,s}\| \geq \varepsilon_t$ With: T_{sa} = Supply air temperature $T_{sa,s}$ = Supply air temperature set point ε_t = threshold parameter accounting for errors in temperature measurement Mode= 1, Operation maximum fan speed (100 %)	Supply air temperature of AHU doesn't reach set point (outside the dead band) by maximum fan speed. Potential causes: Controller not well tuned Capacity of heat exchanger is undersized Hot water supply temperature is too low Not enough cooling capacity in cooling coil Cold water temperature is too high Chiller is off
2	IF Mode = 0: AHU = 1	There is no fit with operational schedule Potential causes: HVAC system is overridden manually Summer and winter months not accounted for Bank holidays not accounted for Recommendations: Check if HVAC system is overridden manually Check software settings related to Winter- and Summer Check holiday settings in the software
3	IF mode = 1: $T_{sa} - T_{sa,max} \geq \varepsilon_t$ With: $T_{sa,max}$ = 20.0 C, T supply air max.	AHU supply air temperature exceeds maximum of 20.0 °C by maximum fan speed
4	IF mode = 1: $T_{ra} - T_{ra,s,max} \geq \varepsilon_t$ With: T_{ra} = exhaust air temperature $T_{ra,s,max}$ = 21.0 C, T exhaust air max.	Exhaust air temperature exceeds its maximum value of 21.0 °C by maximum fan speed
5	IF Mode (0,1): pump twin-coil=0, pump heat exchanger = 1	The heat recovery pump (Twin-coil)must not be OFF while heat exchanger pump is ON
6	IF Mode (0,1): number of ON-OFF transitions per hour $\geq S_{max}$	The system is unstable if the number of ON-OFF transitions per hour $\geq S_{max}$

Results

BAS system monitoring period: Nov. 12, 2006 to Feb. 2, 2007.

Useful measurements : 19.025

Day regime measurements : 7921

Night regime measurements : 11104

Time interval : 5 min (1,585 h, 66 d).

Software platform: spreadsheet

Table 4-13 summarizes the expert rule analysis results. For example, when evaluating the 7921 day regime measurements for case No. 1, the supply temperature was found to not reach the setpoint 177 times, or approximately 2.2 % of the operating period. Similarly, when evaluating the 11104 night regime measurements for case No. 1, the supply temperature was found to not reach the setpoint 6376 times, or approximately 57.4 % of the operating period.

Table 4-13 Expert rules analysis results

No	Test condition	Test condition occurs	Error % Day regime (Fans 100 %)	Error % Night regime (Fans 50 %)
1	Supply air doesn't reach set point.	Fans 100 % (Daily operation): $\varepsilon_t = 1.0$ C, **177** times the supply air exceeds its set point Fans 50 % (Night operation): $\varepsilon_t = 1.0$ C, **6376** times the air supply exceeds its set point	2.2 %	57.4 %
2	AHU test fan speed. Fans run 100 % instead of 50 %.	**734** times (147 h!)	9.3 %	
3	T- supply air is maximum 20 C.	Fans 100 %: T-supply > 20.5 C: **63** times Fans 50 %: T-supply > 20.5 C: **2211** times	0.8 %	19.9 %
4	T- return air is maximum 21.0 C	Fans 100 %: T- return air > 21.5 C: **2,749** times Fans 50 %: T- return air > 21.5 C: **6,878** times	34.7 %	62.0 %
5	Absolute humidity return air doesn't reach set point.	Fans 100 %: Set point +/- 2.0 gram: **1889** times Fans 50 %: Set point +/- 2.0 gram: **1766** times	23.9 %	15.9 %
6	Valve twin-coil not fully open when heat exchanger ON.	This situation reach: **106** times	1.3 %	
7	Switch heat exchanger pump on and off frequently.	Number of times that pump switched on/off < 5 min: **1,322**		16.7 %
8	Switch both chillers on and off frequently.	Re-adjust chiller controllers to avoid frequent switching on-off in very short time.		
9	Control signal of the three-way valve of the cooling coil.	The three-way valve cannot be closed fully. Based on monitoring data, the minimum valve position is 20 %. The valve cannot be checked for the presence of a threshold value or something else, because of other temperature sensors.		

Conclusions based on information from Table 4-13 is unstable control behavior. The table shows that during night regime (50 % air flow) the control of the supply air is unstable. Figure 4-42 illustrates this for a selected week of the supply air set-point and the measured air supply. The figure shows unstable behavior of control during night regime (50 % air flow).

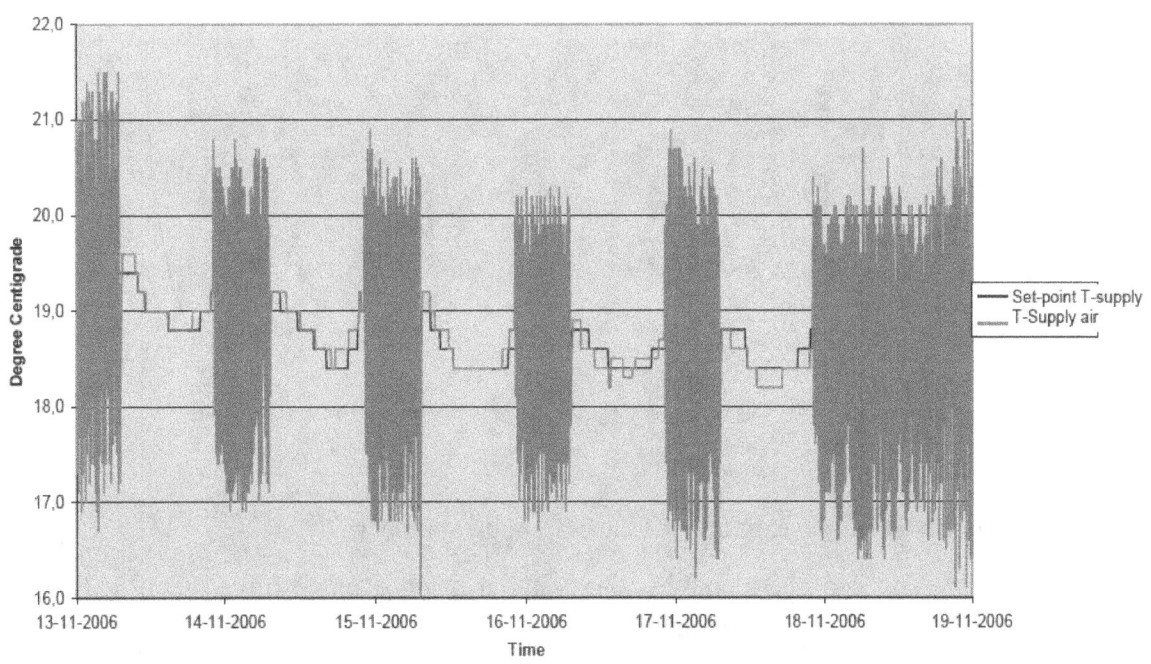

T-supply air Set point / Measured supply air

Figure 4-41   Set-point and temperature of supply air.

The second row shows that for 147 hours fans were working in day regime (100 % air flow) instead of night regime (50 % air flow). An example of this fault is presented in Figure 4-43, which shows the BAS system does not account for holiday periods.

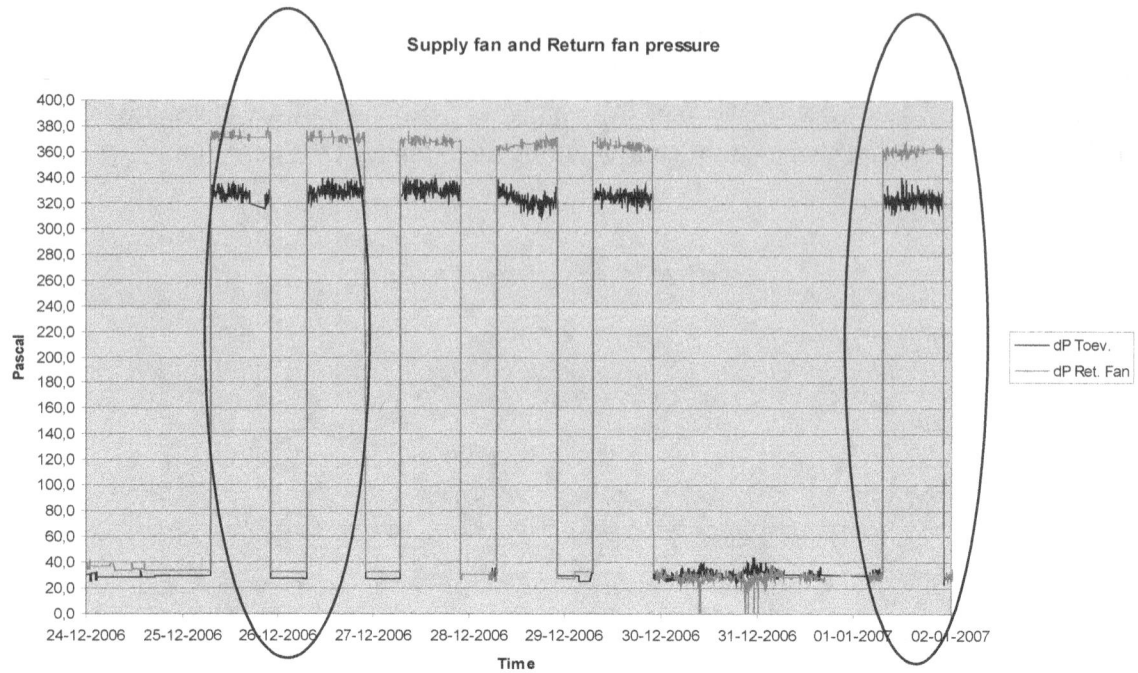

Figure 4-42   **The supply and return fan pressure in day and night regime.**

Conclusions are based on two examples of the bottom up approach.

The first example is based on troubleshooting without using detailed information of the BAS system for analysis. The starting point of this approach is the tenant complaints concerning building comfort. The objective is to reduce the amount of complaints to an acceptable level, which is why functional test are always done in the building and its HVAC system. Additional simple data logging sets are used as a basis for specific data collections for a short period, enabling us to solve the problems.

The second example is a blind analysis of an AHU based on:

- Functional description of the AHU as a basis for deriving expert rules;

- Data collected from a BAS system; and

- No additional sensors being used.

It is possible to blindly investigate AHU performance using information collected by the BAS system. The following information on the HVAC system is required:

- Design and recent modifications;

- Testing, adjusting and balancing reports; and

- Functional description.

The case study is summarized as follows:

- A total of 16 simple expert rules are made from the functional description;
- Faults are detected and solved based on this outcome. The most important detected fault was the sometimes frequent ON/ OFF change of the two chillers. This was why one of the chillers burnt out.
- The starting point for introducing ongoing commissioning is that the basic performance of the HVAC system is OK. If not, Re-commissioning must be executed before starting ongoing commissioning;
- Even if the HVAC systems works under normal circumstances as expected, use of diagnostic tools may reveal imperfections;
- Ongoing commissioning is necessary to monitor the performance of the HVAC system during its whole life cycle;
- Diagnostic instruments support the ongoing commissioning process. They (i) save on maintenance costs, (ii) increase the life span of components, and (iii) save energy.
- There is a perceptible shift in the type of HVAC maintenance contracts. The ordinary maintenance contract based on effort obligation/commitment is changing to maintenance contracts based on performance contracts. In this case, diagnostic instruments will be helpful in clarifying delivered performance;
  Our experience up to now is that after hand over of the HVAC system to the owner/user, they are unwilling to invest in ongoing commissioning.

4.4.3   NEDO Promotion Program (Top-down and Bottom-up), Japan

New Energy and Industrial Technology Development Organization (NEDO), an independent administrative body, is the largest core agency in Japan that promotes R&D and the proliferation of Japanese industrial technology and energy/environmental technology. It offers proliferation programs for the development and introduction of new energy/energy-conservation technology, as well as R&D projects on industrial technology, etc. One of NEDO's grant-aided projects is the Promotion Program for the Introduction of Efficient Energy Systems in Housing/Buildings (to assist with introduction of BEMS). This project offers subsidies to businesses that install BEMS in their buildings and meet the program's criteria, aiming to promote energy conservation under an appropriate control.

Criteria for Subsidy Eligibility:

- The possibility exists for introduction of BEMS to existing, newly-built, extended or rebuilt buildings;
- Energy consumption can potentially be reduced by BEMS;
- Energy performance can be measured by equipment segment such as heat sources (refrigerating machines, heat pumps, cooling towers), pumps, lightings/outlets, and others;

- A well equipped energy management system exists to collect measurement data and store it; and
- To be able to accomplish the grant-aided project, and to report for 3 consecutive years after introduction of BEMS.

The project started in 2002; by 2008 more than 200 businesses had enrolled. The distribution of grant-aided businesses by building size is shown in Figure 4-43 on the left. Large-sized buildings with over 10,000 m^2 make up a greater portion. The breakdown of grant-aided businesses by building usage is shown in Figure 4-43 on the right. Various businesses have enrolled. Among them those using their buildings as offices and stores represent the majority. Figure 4-44 shows the energy-conservation attainment levels of the grant-aided businesses by building usage. All businesses attained higher levels of energy-conservation than the targets set out in the planning stage.

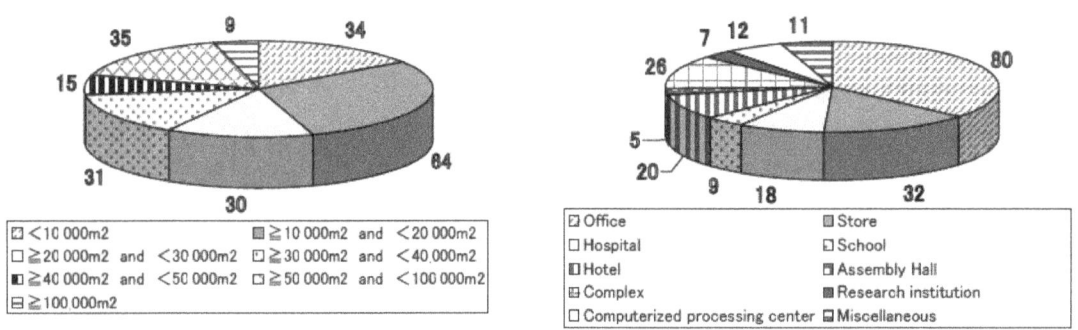

Figure 4-43   (Left): Grant-aided Businesses by Building Size< m^2 (Right): Grant-aided Businesses by Building Usage

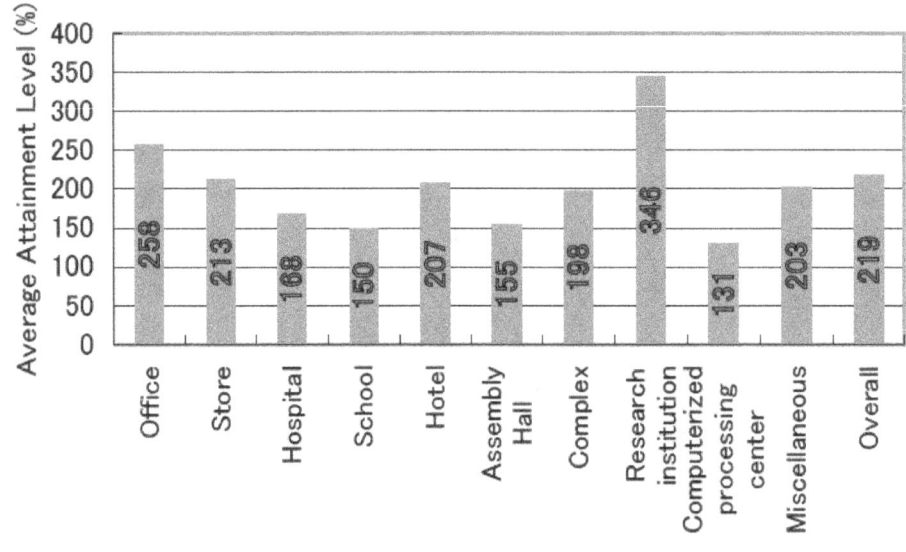

Figure 4-44   Level of energy conservation attained by Building Usage

These are the actual results in fiscal year (FY) 2007 of a total of 212 grant-aided businesses enrolled from FY 2003 to FY 2006:

Total amount of energy-conservation: 2,423 TJ/year
Rate of energy-conservation: 10.1 %
Cost-benefit performance: 91.1 MJ/year/1,000 yen

The reasons all grant-aided businesses attained their energy-conservation targets and derived benefits are:

1. Adequate target values are set

    The grant-aided business reports their energy consumption to NEDO before the introduction of BEMS. Predicted values of energy consumption, as well as grounds for the prediction when introducing BEMS, are examined at the time of inclusion. The appropriateness of the targets is evaluated by a third party.

2. Attainment levels must be measured and reported.

    This program requires participating businesses to measure energy consumption by element; e.g., heat sources, lighting/outlets etc., and report levels of goal attainment based on actual energy consuming performance. If target values are not attained, they must report the reasons to NEDO. This motivates the businesses to make every effort to attain their goals.

3. Energy management system of grant-aided businesses is established.

    For grant-aided businesses to attain their target values of energy-conservation, it is fundamental that their energy management systems be well established. NEDO decides whether or not to include a business by checking their energy management system. The business organizes their energy management system with participation of experts such as designers, instrumentation manufacturers, members of ESCO, etc., regularly checks the status of goal attainment; and, in the case of non-attainment, analyzes the causes and makes necessary improvements.

Figure 4-45 shows the relationship between frequency of utilizing BEMS data and energy-conservation rates in grant-aided businesses. Many businesses use BEMS data once a month. Also, it was confirmed that regularly considering improvement measures in performance confirmation meetings tends to result in higher rates of energy-conservation.

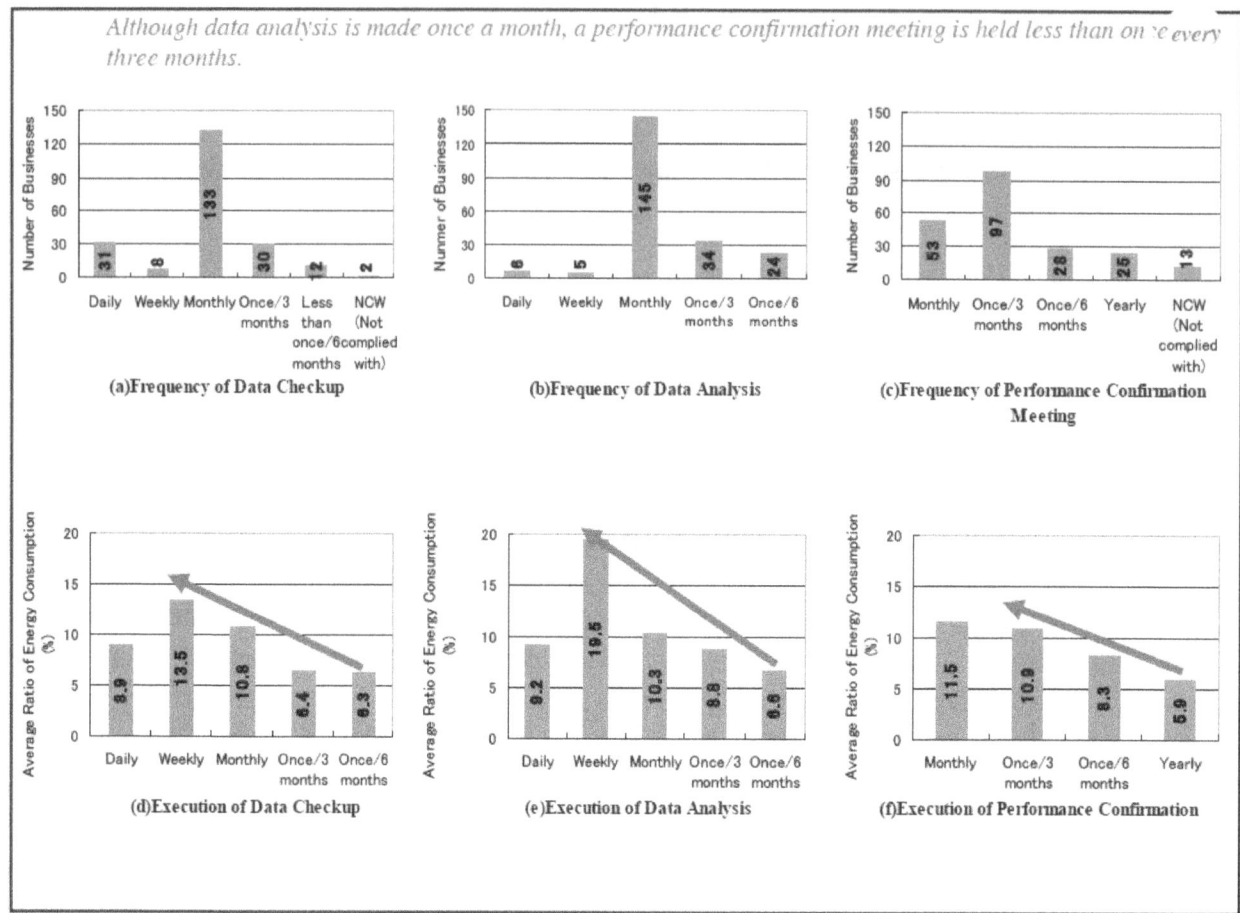

Although data analysis is made once a month, a performance confirmation meeting is held less than on 'e every three months.

(a)Frequency of Data Checkup

(b)Frequency of Data Analysis

(c)Frequency of Performance Confirmation Meeting

(d)Execution of Data Checkup

(e)Execution of Data Analysis

(f)Execution of Performance Confirmation

Figure 4-45　　　Confirmed results on frequency of utilizing BEMS data

The "Commissioning Process" is used to document and verify target fulfillment. Clearly detailed targets are established in writing within the OPR, including quantitative values and conditions for their attainment, and a third-party CA supporting the fulfillment.

Although the Promotion Program for the Introduction of Efficient Energy Systems in Housing/Buildings (to assist with BEMS Introduction) is not planned prior to assumption of such commissioning, in this project, grant-aided businesses specify their demands for achieving energy-conservation in a document in the application form. The implementation report includes information on quantitative target values and conditions for their attainment, how to analyze measured data, solve problems under an energy management system consisting of experts, such as a designer and others, in addition to the building's owner, and the procedure for reporting goal attainment status to NEDO. This process is considered similar to the Commissioning Process.

Deployment of sensors

Grant-aided businesses deploy sensors in their buildings to meet requirements for measurement items, as part of the project criteria and to determine the cost-benefit performance of the energy-conservation strategy, as recommended by NEDO. Figure 4-46 shows the requisite and recommended measurement items.

4. Requisite Measurement Items

The building's energy consumption from heat sources, pumps, lighting/outlets and other lines related to energy consumption volume must be measured and monthly integrated values reported.

5. Recommended Measurement Items

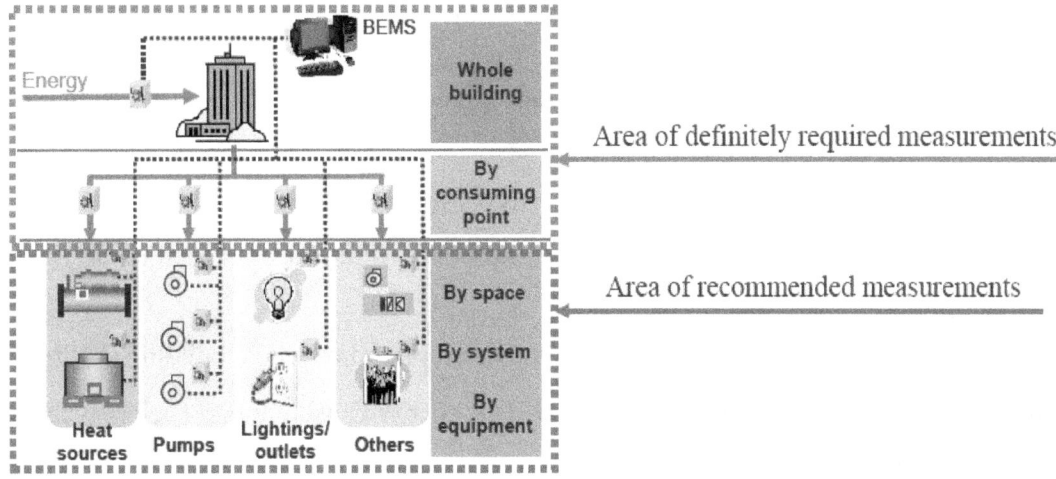

Figure 4-46   Segmented image of sensor deployment in grant-aided project

It is recommended that measurement of each target for each energy-conservation strategy be verified. For example, when introducing a variable temperature control for cooling water in a water-cooled refrigerating machine, the effect of this strategy is verified by measuring the temperature of cooling water and the motive power of the refrigerating machine, the cooling tower fan and the cooling water pump.

Performance Metric

The performance metric is divided into two classifications: the mandatory one required for reports to NEDO and the one each business independently defines to promote energy-conservation.

1. Mandatory metric required for NEDO reports

A business is required to annually report that their planned value for the rate of energy-conservation in the whole building is attained for the three years following the project's implementation. Therefore, the rate of energy-conservation in a whole building is mandatory as a performance metric.

103

A planned value is found using Formula (1); and an actual performance value using Formula (2). If events alter the planned value, e.g., changes to operational conditions, etc., after the planning stage, modifications must be made to redefine targets and specifications.

Rate of Energy-conservation (Planned Value) [%] = (E_{ref} – E_{pre}) ÷ E_{ref} ×100:   *Formula (1)*

Rate of Energy-conservation (Actual Value) [%] = (E_{ref} – E) ÷ E_{ref} ×100:   *Formula (2)*

Note:

E_{ref}: Standard annual energy consumption of a whole building (MJ/Year)

For new construction:

To be calculated in accordance with the Coefficient of Energy Consumption (CEC). The value should be set by predicting any effects on the energy-conservation approach exempt from the grant-aided project.

For existing buildings:

To be calculated from a mean value of actual performance in the last three years.

E_{pre}: Planned annual energy consumption of a whole building (MJ/Year)

A value taking into account the effects of energy-conservation strategies included in the grant-aided project.

E: Actual performance value of annual energy consumption in a whole building (MJ/Year)

CEC is an index related to the equipments' established energy consumption efficiency based on Japan's "Regulation on rationalization of the use of energy." It is used as a criterion of energy conservation performance in building in Japan, and it shows energy conservation in the smallest value. The criteria value is set for air conditioning facilities, ventilation equipment, lighting equipment, equipment for supplying hot water, and the elevator equipment depending on the building's usage. There is an obligation in Japan to report to the government when building an architectural area in excess of 2 000 m², or when doing large-scale repair work. A rule for the relative calculation of the index value is provided.

CEC = Energy consumed by object equipment ÷ Reference value of energy consumed by object equipment

2. Metric independently established by each business for promoting energy-conservation

The research results on metrics that each business independently establishes to promote their energy-conservation project are shown in Figure 4-47. Many businesses include a building's overall power consumption volume, overall cold energy consumption volume and overall heat consumption volume, as well as monitoring information of indoor conditions such as temperature and humidity. It is assumed they use a top-down approach. Meanwhile, some businesses adopt metrics to evaluate equipment and systems such as usage-based temperature differences and unit COP of heat source equipment. Some businesses may adopt a bottom-up approach.

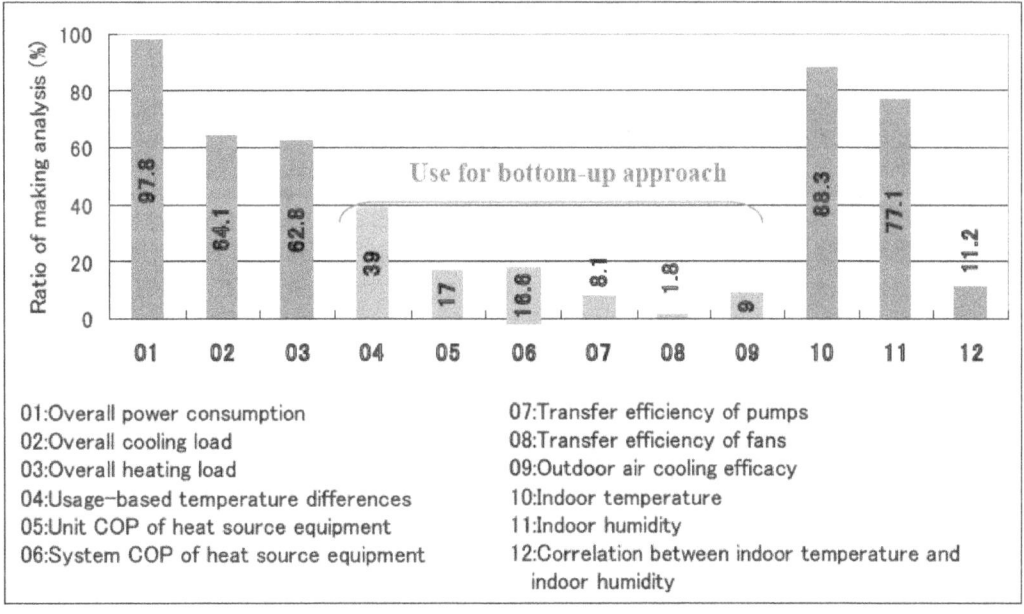

01:Overall power consumption
02:Overall cooling load
03:Overall heating load
04:Usage–based temperature differences
05:Unit COP of heat source equipment
06:System COP of heat source equipment

07:Transfer efficiency of pumps
08:Transfer efficiency of fans
09:Outdoor air cooling efficacy
10:Indoor temperature
11:Indoor humidity
12:Correlation between indoor temperature and indoor humidity

Figure 4-47:   Adoption rate of performance metric by grant-aided businesses

Building case study 1: Businesses using top-down approach

Research on businesses adopting a top-down approach shows that they check monthly integrated values for energy consumption in a whole building or by consuming point and the comparison ratio of annual energy consumption volume by consuming point to the previous fiscal year, etc. (see Figure 4-48 and Figure 4-49). For improvement steps, many businesses identify the month and the consuming points that show increased energy consumption in an energy-conservation confirmation meeting, then discuss possible causes for the increase with those involved, and make on-site confirmation(See Figure 4-51).

For example, overall energy consumption volume in the building is higher than the planned value. The energy consumption volume in the lighting/outlet lines is higher compared to the previous year. This is identified through deliberation in an energy-conservation confirmation meeting and confirmed by checking if lights are not being turned off or the lighting-up status(See Figure 4-49).

The attainment level of the energy-conservation project tends to depend on the level of expertise involved in performance confirmation meetings.

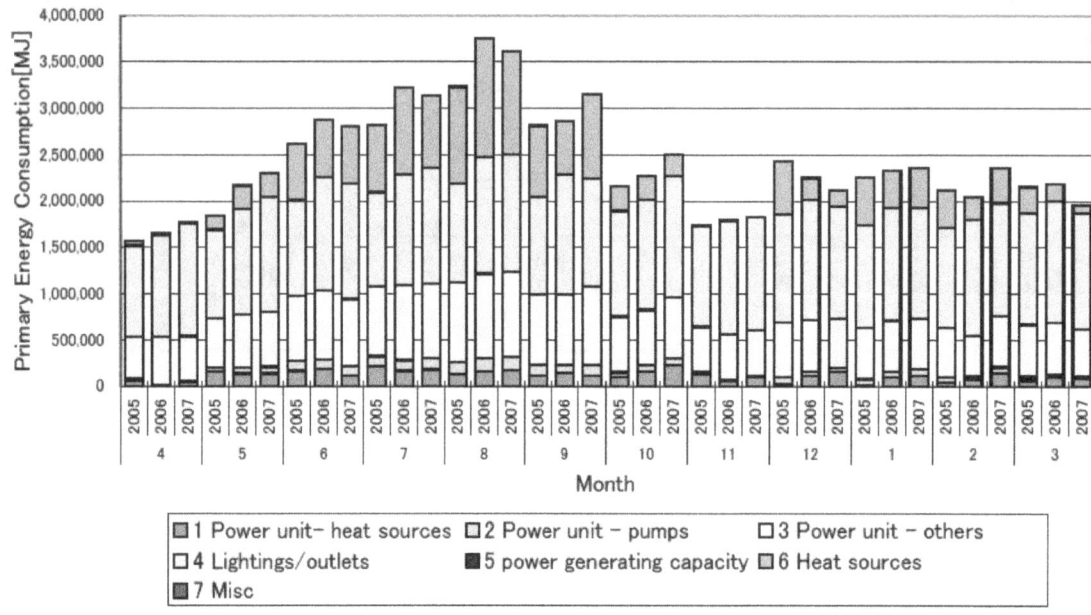

Figure 4-48   Comparison of annual energy consumption by consuming point between a target year and previous year

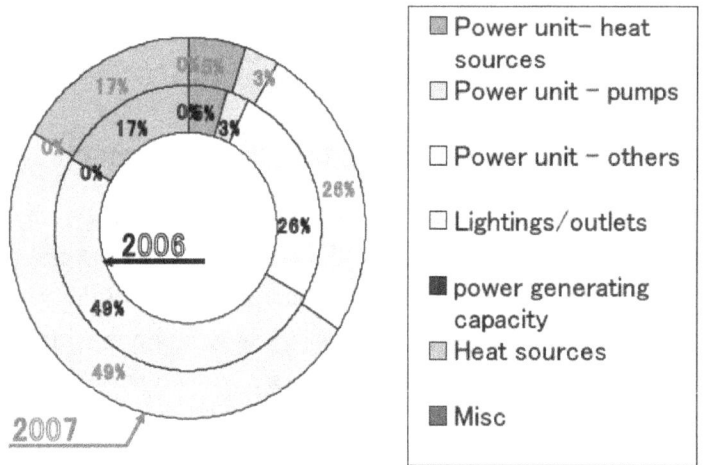

Figure 4-49   Biographic energy data comparison enables confirmation of attainment of conservation goals

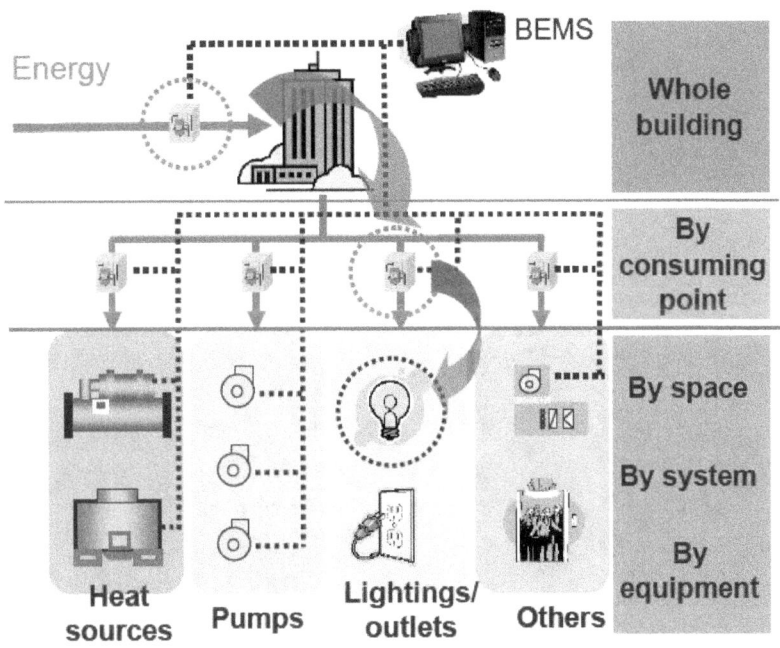

Figure 4-50   Steps for improving performance by businesses using top-down approach

Building case study 2: Businesses using bottom-up approach

Figure 4-51 shows a grant-aided business adopting a variable flow control of cold-water pumps. It is a plotted display of the measurement values in a year: the horizontal axis shows calorimetric values for water conveyance rates of cold-water pumps and the vertical axis shows the motive power of pumps. The higher a value is plotted, the lower its efficiency. Trendlines for low-level pumps are plotted in a higher position than hig-level pumps, showing they have lower efficiency. This suggests that values of pressure set for low-level pumps are relatively high and leave room for readjustment. This way, the grant-aided businesses adopting bottom-up approaches prepare performance metrics and graphs for each method of energy-conservation (see Figure 4-52) and make improvements using expert opinions from performance confirmation meetings. Operators involved in performance confirmation meetings can learn how to assess conditions, and rational reinforcement of the management system can be anticipated. In addition, those who adopt bottom-up approaches use top-down approaches concurrently, and have a greater tendency to show high level goal attainment than those not adopting bottom-up approaches.

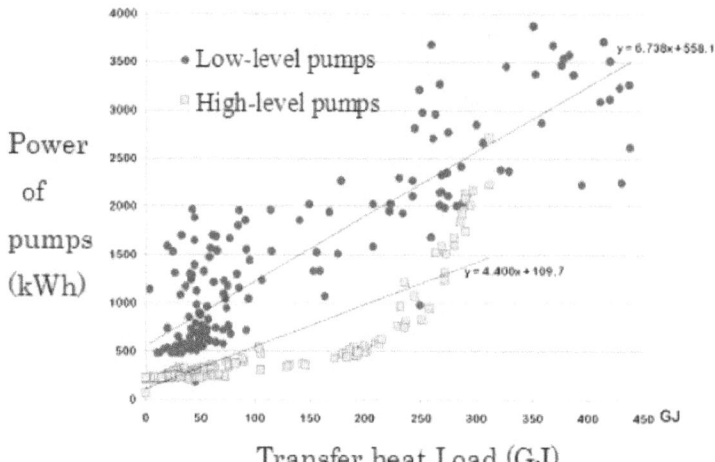

Figure 4-51   **Improving Performance by adopting a bottom-up approach**

Examples of the bottom up approach using a graph of BEMS data include:

1) Comparing actual energy consumption in a whole building with the planned value, and then to judge the priority for conducting improvements.

2) Determining energy consumption for heat source and pump with compared values from the previous fiscal year, or others.

3) Checking the efficiency of refrigerating machines and the operating condition of variable flow pumps by making a graph of BEMS data, and to judge presence or absence of problems.

4) Determining a course of action through deliberation in an energy-conservation confirmation meeting.

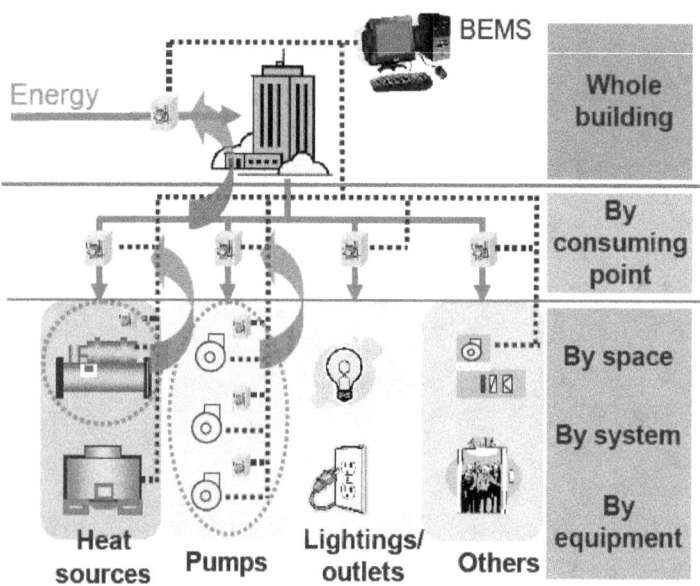

Figure 4-52 Improving performance of businesses using bottom-up approach

4.4.4 SHASE Energy Performance Measurement Manual

In Japan, while awareness of the need for energy-conservation is rising, there are few buildings in which performance of the architectural equipment is understood. Even though the performance is grasped, the methods for understanding differ with buildings; the absence of uniform standards is seen as problematic. In an aim at solving this problem, SHASE prepared the "Energy Performance Measurement Manual for Building Equipment & Systems." This manual organizes views on performance grasping methods for air-conditioning facilities and plumbing sanitary systems, and also shows practically available measurement methods and how to process data obtained with these methods, by equipment & system. In addition, it covers specifications regarded as common elements of measurement and how to use measuring equipment available for the measurement. Moreover, this manual presents a point of view for grading of measurement in accordance with purposes for obtaining data, points to consider in electrical measurement inevitable to the grasping of energy consumption, and some examples of measurement/analysis for several real buildings.

Utilization of this manual makes it possible to make analyses by specifying measurement points and evaluation indexes for performance grasping at levels of system and equipment. It is being used as a bottom-up approach. The table of contents of this manual is shown in Table 4-14.

How to Use the Manual

When designing an actual measurement plan, one first envisages performance items desired to grasp/evaluate in reference to the analysis examples illustrated in Chapters 3 and 4, and then chooses numerical evaluation items to indicate the performance. Secondly, one chooses system diagrams and standard measurement points and configures these on actual equipment and systems. Meanwhile, there are different groups of evaluation items such as component, sub-system, and whole system, etc. according to the range of performance grasping. Since expenses and labor hours differ widely according to fineness of measurement contents, measuring period and measuring equipment used, three levels of measurement from Level 3 (minute measurement) through Level 1(standard measurement) are set according to specifications of measurement. Consideration is given so that one can conduct well-balanced and complete measurement by taking the difference of levels into account when designing a measurement plan. In addition, for instructions on selecting and installing measuring equipment, refer to the items for individual equipment in Chapter 6.

Table 4-14 Table of Contents of SHASE Energy Performance Measurement Manual for Equipment & Systems

1		Objective and general configuration
2		Basic view to evaluate energy performance
3		Energy performance grasping for air-conditioning facilities
	3.1	Heat source systems
	3.2	Water conveyance systems (after secondary pumps)

Measuring the usage of electricity by electrical system may enable efficient grasping of energy performance. For this, giving consideration at the design stage of electrical systems is effective. Several examples of related recommended system design are presented in Chapter 5.

To add to the convenience of this manual's usage on site or for batch grasping an entire picture in brief at the design stage, the "Performance Management/Energy Management Sheet," is found at the end of the manual as Attachment 1. It includes measurement items, levels, outline of evaluation items, numerical processing, graphic representations, evaluation methods and feed-back methods for measuring bracketed results.

For the management and handling of individual measurement points and data, a uniform rule for identifying data is recommended. This manual suggests the use of \ naming codes that reflect the type of target equipment and data contents, etc. as specific naming conventions. Attachment 2 provides naming rules and a list of codes.

A CD recording raw data of measurements in actual buildings on which evaluation results shown in this manual are based is provided as Attachment 3. It demonstrates an analysis for each item evaluated and the plotting process.

Sensor Deployment

The manual provides uniform standards for the planning of sensor deployment, as follows:

- Unification of target range for performance evaluation

For optimum understanding of architectural equipment systems, it is best to evaluate the validity of the target systems' performance by referring to performance in buildings similar to that of the target building. If the range of target systems for evaluation differs between buildings, however, it will be impossible to make equal comparisons. For instance, with the heat source system shown in Figure 4-53, secondary pumps may be included in the range of heat source system. In this manual, the target range for each system is unified, and clearly defined through mapping.

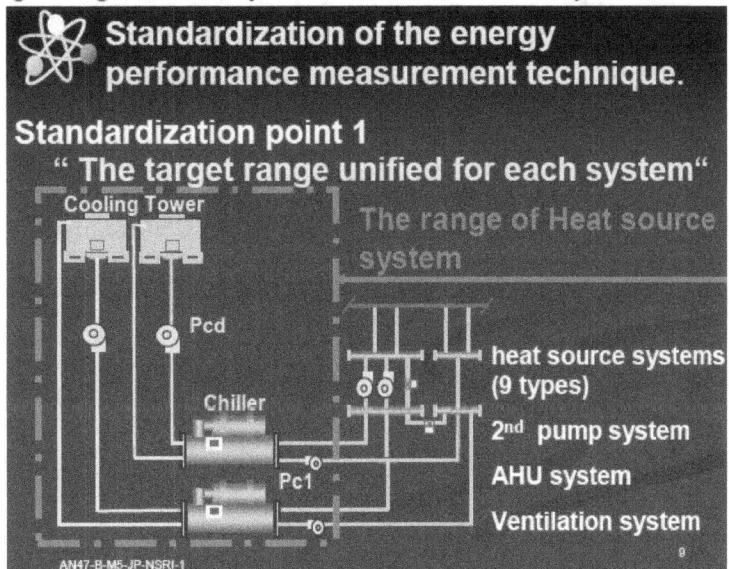

Figure 4-53 Unified image of target range for performance evaluation

111

- Unified location for installing and naming measurement points

Even though the same evaluation indexes are used, numerical values will have different significance if the measurement point concept used for calculations differs. Also, as names of measurement points vary between buildings, the time required for a third party to grasp measurement data is lengthy. This manual sets out unified standards for naming and locating measurement points for architectural equipment. TSC21 naming code is used to name points (See Figure 4-54).

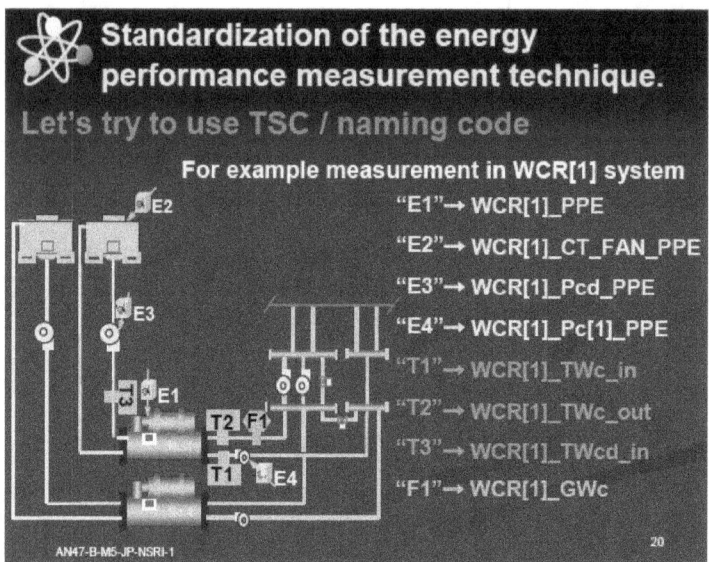

Figure 4-54   Overview of measurement point names

When installing measurement points, it is crucial to prepare a plan so that performance may be grasped in the early stages of design and construction, for purposes of consistency, accuracy, and to further reduce initial costs. Minimum measurement points must be specified along with clearly defined objectives for conducting performance evaluation. Figure 4-55 shows the evaluation indexes and measurement points required to conduct the performance evaluation of a heat source system, including a water-cooling electric chiller. When the chiller's unit COP is chosen as an evaluation index with which to evaluate the energy performance of a heat source system, it is fundamental to measure the temperature at the cold water inlet and outlet, the flow rate of cold water and the chiller's electric power consumption. The temperature at the cooling water inlet should also be measured for comparison with the performance curve provided by the manufacturer.

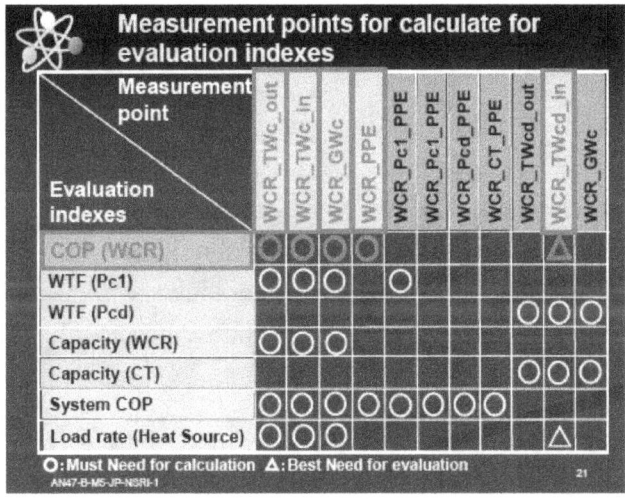

Figure 4-55   Correlation between evaluation indexes and measurement points

- Designing reasonable measurement plans

The desired level of measurement accuracy is based on the measurement equipment/system, recording technique (media), recording interval and period, etc. Table 4-15 shows the recommended accuracy levels by item. The example provided in Table 4-16 shows levels at typical measurement points of a heat storage air-conditioning system. Table 4-17 shows standard measuring instruments. When assessing performance in a heat storage air-conditioning system, focus should be placed on energy performance and the status of nighttime transition of load. Particularly accurate measurements are essential for thermal quantity (flow rate/temperature), electric power consumption and start-stop information. Use of Measurement Level 1 is ruled out because of its inability to accurately measure integrated thermal quantity. For systems whose calculations/evaluations are made using integrated thermal quantity data such as heat source equipment COP, thermal storage efficiency and thermal storage tank utilization factors, etc., the measuring plan or specifications should stipulate the use of Levels 2 or 3 measurement.

Many types of automatic recoding instruments have recently become available at low prices; temporary use of Level 2 should be done at the acceptance inspection. Even if a central monitoring system is well established, close attention should be paid since recording on magnetic media may leak from the original specification.

Although calorimeter usage is common when measuring thermal quantity, as flow rate and temperature data are required for any performance evaluation, it may become necessary to set up electromagnetic flow-meters when DDC is not used as a control system, or when measurement signals from electromagnetic meters for existing control systems cannot be removed. Calculating thermal quantity (both instantaneous and integrated values) and processing primary data are done more efficiently with regard to cost and data confirmation using functions on a measuring PC or central monitoring system.

In some cases, 1-minute measurement/recording intervals may be necessary for an operational acceptance inspection at the component level. When conducting performance evaluations, the

113

large amount of data may pose an obstacle to data operation/analysis. Therefore, it is desirable to make the recording intervals variable.

For basic performance evaluation the recording interval should be set at 15 minutes. If set at less than 15 minutes at Levels 2 or 3, a mean value should be recorded automatically every 15 minutes. If this is impossible, automatic recordings should be made in the maximum assignable recoding interval; later the mean value for 15 minutes should be calculated at the analysis stage. In addition, if thermal quantities are calculated in the analysis stage, the recording interval for temperature and flow rate should be pre-set to the same length as the thermal quantity measurement interval. .

Table 4-15   Measurement levels

	Measuring equipment	Recording method	Recording interval	Measurement interval
Level 3 (detail, Expensive)	Electronic detector, Data logger, Central Monitoring System	Automatic recording by electromagnetic media	Instantaneous value – 15 min Integrated value – 1 h	Instantaneous value – less than 15 min Integrated value – 1 h
Level 2 (moderate)	Temporary electronic detector, Data logger	Automatic recording by log sheets or Electromagnetic media	Ditto	Ditto
Level 1 (simple, Inexpensive)	Measuring instruments fixed on site	Manual recording by visual inspection	1 h	1 h

Table 4-16 Target Measurement Points and Measurement Level

	Measurement items							
	Flow rate	Integrated quantity of water	Pressure of cold/hot water	Water temperature	Current value	Integral power consumption	Electric power	Start-stop state
Heat pump	3, 2		2, 1	3, 2	2, 1	3, 2		3, 2
Chiller	3, 2		2, 1	3, 2	2, 1	3, 2		3, 2
Cooling tower	2, 1		2, 1	2, 1	2, 1	3, 2		3, 2
Condensing Water pump	2, 1		2, 1	2, 1	2, 1	3, 2		3, 2
Primary pump	3, 2		2, 1	3, 2	2, 1	3, 2		3, 2
Secondary pump	3, 2		2, 1	3, 2	2, 1			3, 2
Thermal storage		2, 1		3, 2	2, 1			
Power receiver/ Transformer						3, 2	3, 2	

Table 4-17 Standard Measuring Instruments

Measurement items	Level 3, 2	Level 1
Flow rate	Ultrasonic flowmeter, Electromagnetic flowmeter	Instant flowmeter (area or pitot-tube type)
Integrated water quantity	Integrating flowmeter with pulse transmitter	Water meter
Water pressure	Pressure transmitter	Bourdon tube pressure gauge
Water temperature	Platinum resistance temperature detector, Thermocouple	Glass stick thermometer Round-shape thermometer
Current value	Alternating current transducer	Alternating current meter
Integral power Consumption	Voltmeter with transmitting equipment	Voltmeter
Electric power	Electric power transducer	Electric power meter
Start-stop state	No-voltage a junction, Constant voltage Generator	Visual inspection

Performance Metrics

• Unification of Evaluation Index for Performance Evaluation

Although various evaluation indexes are used to evaluate the performance of architectural equipment, the fact is that different definitional equations are used for similarly named evaluation indexes. This manual specifies major evaluation indexes and their respective architectural equipment, clarifies their definitional equations, and categorizes them for use as uniform standards. Figure 4-56 shows the COP in an electric water-cooled chiller, and Figure 4-57 shows the COP of a heat source system. As understanding the operating condition is a prerequisite to doing an energy performance evaluation, the manual defines the respective evaluation indexes (Figure 4-58).

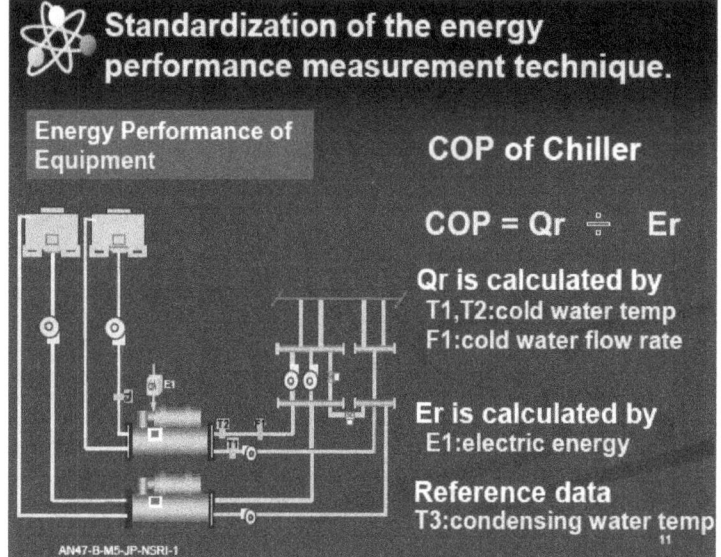

Figure 4-56   Chiller Unit's COP

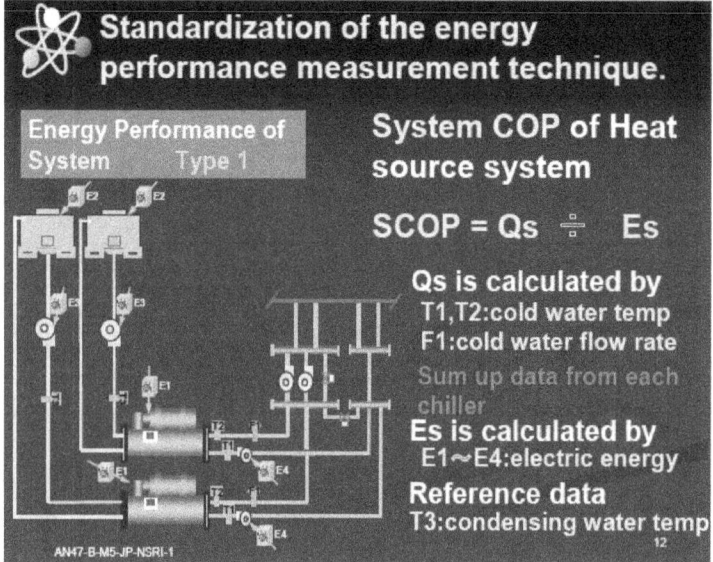

Figure 4-57   Evaluation Image of Heat Source System COP

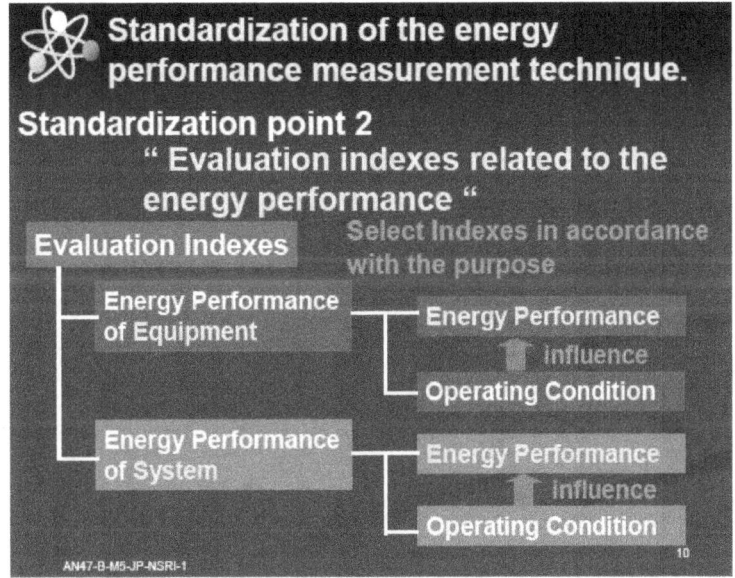

Figure 4-58   Configuration of Evaluation Indexes

- Criteria for Performance Evaluation

As a prerequisite to an evaluation based on performance metrics, measurement data must be validated. Although information from a sensor is generally regarded as correct, attention is required as, sensors can be defective. The followings points should be considered:

a) Basic Check List

Validate absolute values by defining an absolute range.
Check characteristic values by cross-checking characteristic formula of equipment.

b) Validity Check List
To check if equipment is in a valid status (e.g., to neglect temperature and flow rate while the equipment is not in operation, etc.)

c) Cross-checking
To check errors, etc. in data correlation. (e.g., during cooling operation of a heat pump, water temperature at the outlet is 5 °C lower than that at the inlet, etc.)

d) Trend Checking
To check the status of temporal variations (trend). Express multiple related data in the same graph to make efficient and effective checking possible.

These checks should be done in the early stages of measuring works. When invalid data are found: check power source, recalibrate measuring equipment and check measuring equipment wiring immediately. Although the calibration of permanent measuring equipment is done at the construction stage, prior display confirmation is recommended to avoid future problems.

[Example 1]

Although the recording accuracy of temperature is within ±0.5 °C, the temperature difference has a maximum error range of ±1.0 °C. If the temperature range difference of cold/hot water is Δt=5 °C in the design, the difference of temperature/thermal quantity can reach ±20 %. Even if the recording accuracy of temperatures of outward and returning water is satisfactory, necessary corrections should be made as required during data analysis, since any variance can have a profound impact on the difference in temperature/thermal quantity.

[Example 2]

An electromagnetic flow-meter combines detection and computing features; which are pre-calibrated by the manufacturer. There are multiple flow-meters of the same size; altering the detection or computing features after delivery can compromise accuracy. As well as historical management, special attention is required since defects may not be found out unless data is carefully checked during actual operations.

Setting Data Recording Intervals

If the recording interval for data measurement (primary data) is less than 15 minutes, the mean value for 15 minutes should still be calculated. Experience shows that a plot based on 15-minute intervals is best for evaluating daily trends. Shorter intervals for analysis purposes waste time when producing graphs etc.

Calculating Integrated Thermal Quantity

When measuring temperature and instantaneous flow rate without a calorimeter, integrated thermal quantities should be calculated using a personal computer. If the data recording interval is less than 15 minutes, an instantaneous thermal quantity should be calculated based on the primary data, then integrated, before performing the averaging described in Article 2 of the SHASE manual).

Graphing

It is important to improve the visualization of measurement data by graphing. The manual offers information on graphs and how best to interpret them for grasping the performance of respective equipment and systems.

Building case study 3: Usage with an existing building – Shinkawa Project

This example, in which the sensor is installed into an existing building, confirms the relevance of the SHASE manual.

- Outline of Shinkawa building

The Shinkawa office building has a total area of 5 940 m^2. Figure 4-59 outlines the building and its equipment. The HVAC system was recommissioned to an air-source heat pump and ice storage system, and planned using the SHASE manual. The HVAC system is shown in Figure 4-60.

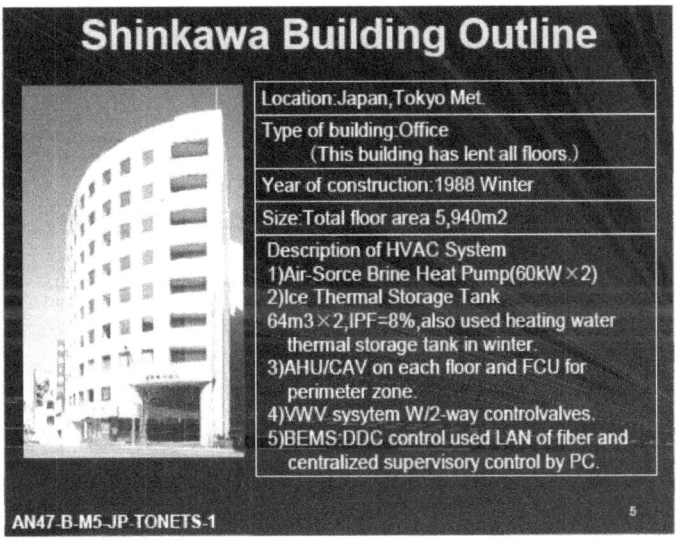

Figure 4-59   Outline of Shinkawa building

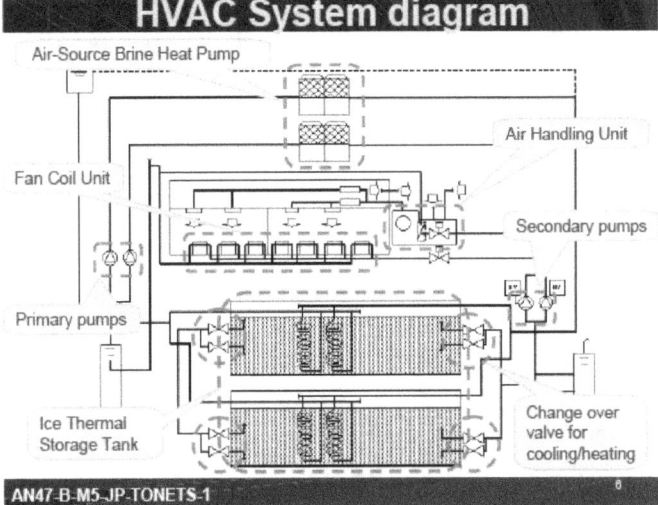

Figure 4-60   HVAC System diagram measurement points

Kind of measurement data

■ The number of measurement

Total number of measurement data				
638				
status	safty	security	alarm	setpoint value
145	40	56	29	34
temperature	humidity	Air/Water flow	calorimeter	power
158	11	30	27	53
pressure	solar radiation	others		
12	3	40		

The number of measurement points is relatively large compared to this size building!

These measurement points were selected to evaluate the energy performance of this building and the HVAC system.

AN47-B-M5-JP-TONETS-1

Figure 4-61   **Number of measurement points**

- Type of measurement data

The measurement data shown in Figure 4-61 reveals a number of measurement points that is relatively large when compared to the building size. These measurement points were selected to evaluate energy performance and HVAC system according to SHASE manual standards.

All of these measurement points are measured by BEMS. The data is used to evaluate everyday operational management and performance.

- Sensor deployment

Sensor deployment and measurement points are shown in Figure 4-62. A comparison between the building's measurement points (red areas) and those of the SHASE manual (green areas) is illustrated in Figure 4-63. Most measurement points correspond to points in the SHASE manual. In this building, sensor deployment can result in the most detailed performance check (Level 3) specified in the SHASE manual. It is too expensive to add the sensor which later ran short to evaluate the fault cause and attain operational rationalization. Even if it is used with an inexpensive temporary measuring device, the problem with sensor accuracy occurs. The measurement plan should be done at the construction stage. It is important to deploy sensors for evaluation and verification that may be needed in future operation phases. This is an important aspect of Commissioning.

The division of levels discussed in the SHASE manual is:

Level 1: To grasp overall building characteristics.

Level 2: To grasp system characteristics.

Level 3: To grasp equipment characteristics

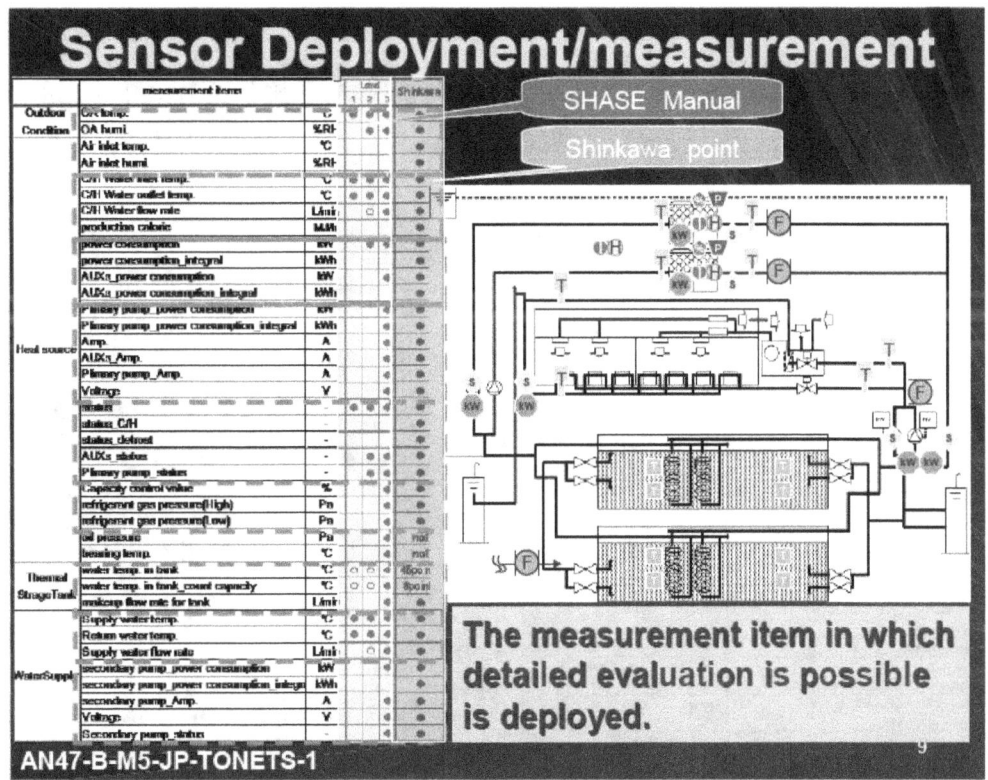

Figure 4-62   Sensor deployment and measurement points

Relationship between evaluated performance and measurement points

Figure 4-63 shows the relationship between evaluation and measurement points.

This measurement point is sufficient to evaluate the energy performance of the HVAC system. The practical methods for data measurement and evaluation practiced as part of the three stages in the bottom-up-approach are explained in Figure 4-63. Step 1 is to check operational status to see if any unnecessary operation is being performed. Its settings are checked against the operation plan, or checked to see if preset values are valid. In Step 2, effective performance of equipment is checked to see whether it conforms to the measurement data for each component. Step 3 is an energy evaluation to ascertain the energy efficiency and load shifting ratio of the ice thermal storage system. It is checked for HVAC system performance and whole building energy consumption.

In this evaluation procedure, a component is evaluated first, then a subsystem, and, finally a whole system.

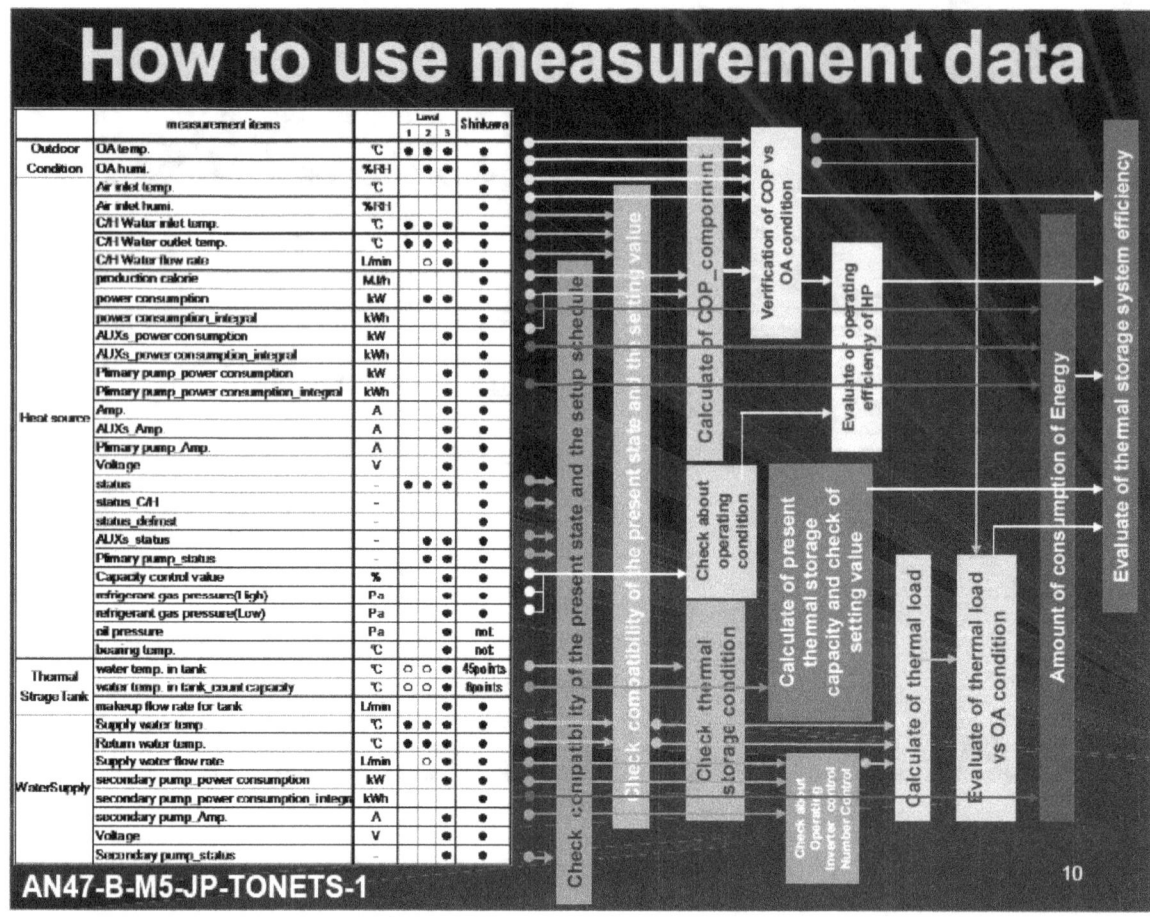

Figure 4-63   Relationship between evaluated performance and measurement points

Evaluation of energy performance that uses measurement data

a . Energy performance of Air source heat pump

Evaluation of COP for air source heat pump is shown in Figure 4-64. The horizontal axis shows the outside temperature; the vertical axis shows COP. It can evaluate if the operating performance is correct by comparing driven COP with the reference value according to outside temperature.

b . Energy performance of secondary pump

Improved control of the number of secondary pumps is shown in Figure 4-65. The horizontal axis shows the water flow rate and the vertical axis shows Pump power consumption. The top figure shows that the data indicates energy is being wasted because the number of operating pumps is too high. The bottom figure shows improved operating conditions.

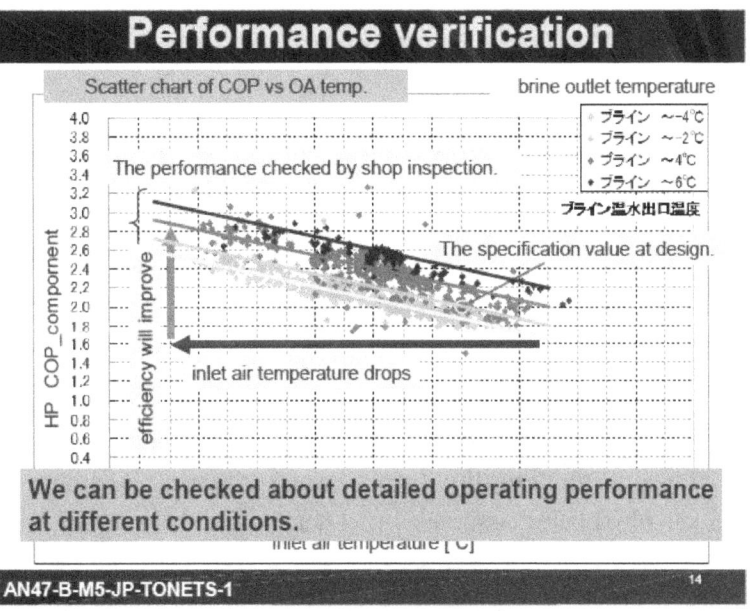

Figure 4-64   Evaluation screen of COP of air source heat pump

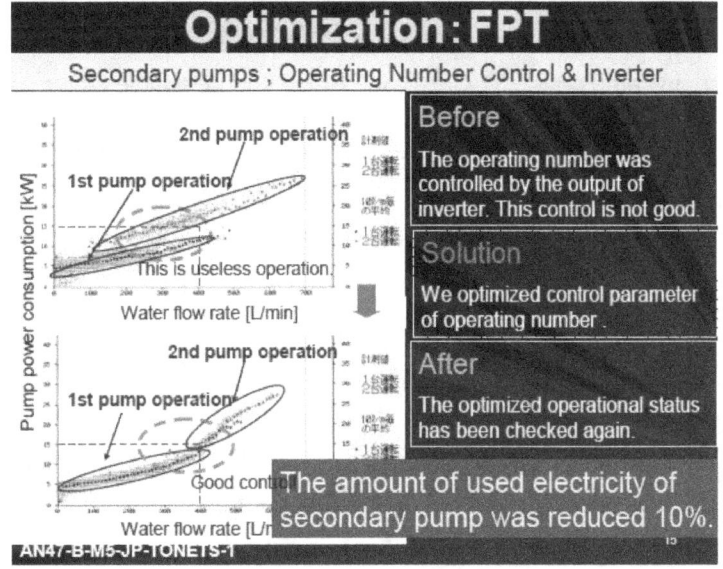

Figure 4-65   Improved control of the number of secondary pumps

Conclusions

The Shinkawa project uses an evaluation technique from the bottom-up approach described in the manual. Advance troubleshooting consists of determining the performance to be evaluated prior to installing the sensors. By performing this evaluation periodically, optimal operation is attained and maintained. This confirms the effectiveness of the SHASE manual.

4.4.5 DABO™ a BEMS-Assisted Commissioning Tool, Canada (Bottom-Up Approach Tool)

Since 1998, the CANMET Energy Building in Varennes (Canada) has been used to test and demonstrate various tools developed in the context of the CANMET Intelligent Building Operating Technologies R&D plan [40], of which DABO is the central component. In this context and for the last eight years an ongoing commissioning process has been conducted. DABO, a software package that uses a hybrid technology composed of conventional and artificial intelligence techniques to ensure optimum operation of building systems has been used actively in the project delivery system for the continuous monitoring of all HVAC equipment and meters (e.g., terminal unit, air handling unit, plant equipment and energy meters), the analysis of the incoming information, the detection and diagnosis of major HVAC component faults, non optimum set points and sequences of operation and the monitoring of implemented measures .

The benefits obtained through the use of the DABO^{TM} software are listed here:

Energy savings	Improved performance of mechanical systems
Reduced greenhouse gas emissions	Rapid and automatic fault detection
Improved occupant comfort	Energy performance monitoring
Capabilities for troubleshooting of malfunctions	Reduced operating costs

General approach:

DABO includes two parts: the commissioning module and a report generator.

The commissioning module, designed to assist and perform some functions described in the ongoing commissioning process section is a module of DABO, which serves as the interface between the end-user (e.g., building operator, commissioning agent, and energy manager) and the control system (BEMS) [41]. As shown in Figure 4-67, the tool continuously monitors the building control data and stores it in a structured database to be used on-line or upon request. Data resulting from standardized test procedures invoked manually or automatically are also stored in the database. The database functions as a server for reasoning algorithms that perform intelligent analyses of the monitored data, performs additional automated tests of components and systems, identifies faults and diagnoses them, and evaluates potential improvements in energy efficiency. The tool produces reports adapted to the different partners involved in the ongoing commissioning process (building operators, service technicians, energy managers, commissioning agents, HVAC&R engineers).

Deployment of sensors:

DABO uses various data values coming from the building energy management system (BEMS). This normally includes all control inputs, outputs, setpoints, schedules, alarms and timers. The performance of DABO depends on the quality and variety of the data supplied to its artificial intelligence engine.

Performance Metric:

Standardized test procedures are performed at three levels using a bottom-up approach. At the first level, an hourly component analysis of individual HVAC devices and equipment is performed automatically using a combination of control loop indices and expert rules to verify their proper operation.

The second level of testing consists of an integrated system analysis to verify the operation and energy performance of the overall HVAC system over a longer period of time (e.g., hours, days, weeks or months). At this level a set of component performance indices and expert rules is also used in the analysis.

The third level performs basic energy performance and operation control quality reports that provide the information required to evaluate potential energy measures on specific devices. Specific applications of the fault detection and diagnostic (FDD) methods implemented in DABO are described further in Section C of the IEA Annex 34 final report [42].

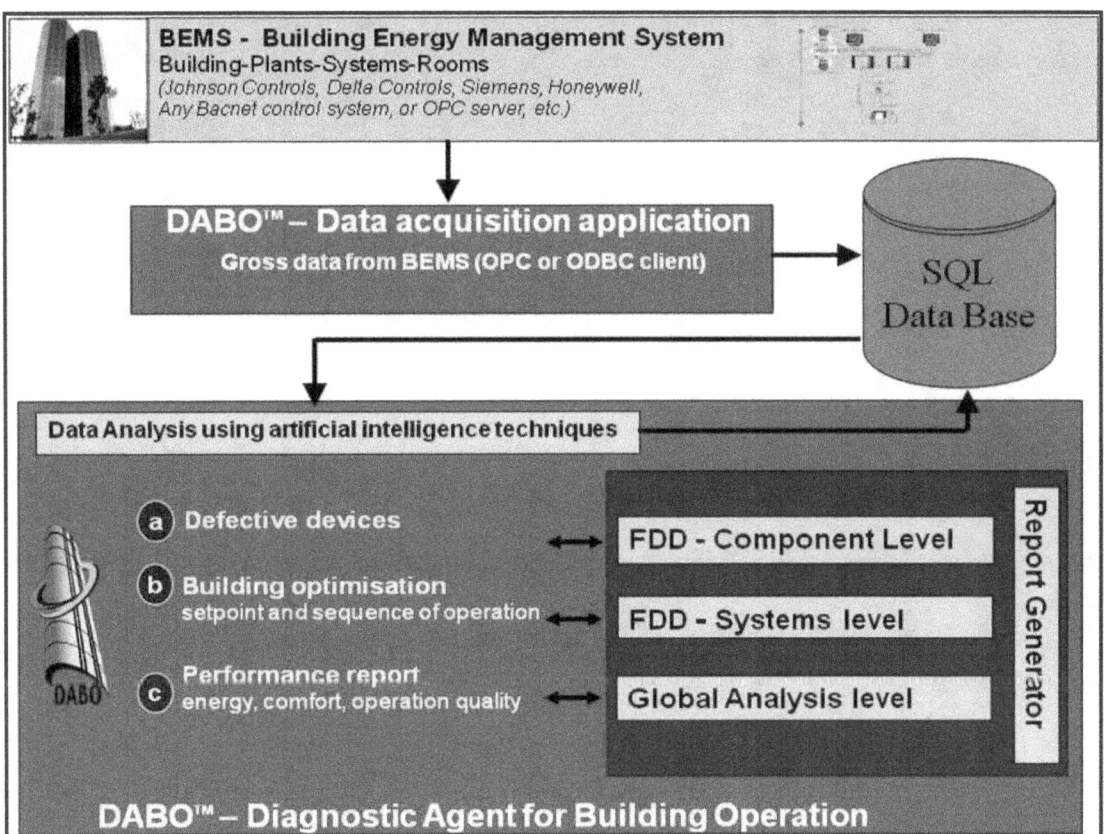

Figure 4-66 Ongoing commissioning tool for HVAC systems embedded in DABO

DABO™ produces four types of report to assist the user in assessing the performance of the building, its systems and devices. These reports primarily display the data in a graphical format and have been designed to allow the most efficient review and use of the information [43]. The reports are summarized in the following list and then presented in further detail.

- Fault Detection and Diagnosis Reports
- Points Reporter
- Commissioning Reports (COMM Reports)
- Fault Management Reports

Fault Detection and Diagnosis Reports (FDD Reports)

These reports display data generated by the various fault detection functions (these are defined as «services» in the software). Data are shown in a color-coded table. Figure 4-67 shows the DABO™ FDD Report (no. 1) with the list of symptoms detected for a single cell (no. 2), and for a single symptom with the related individual graph (no. 3).

The FDD report displays the number of detected symptoms of a failure by date and hour-by-hour for all of a building's air handling, air terminal devices and control zones. It shows also the operation mode and provides access to detailed reports (available by clicking on the + button).

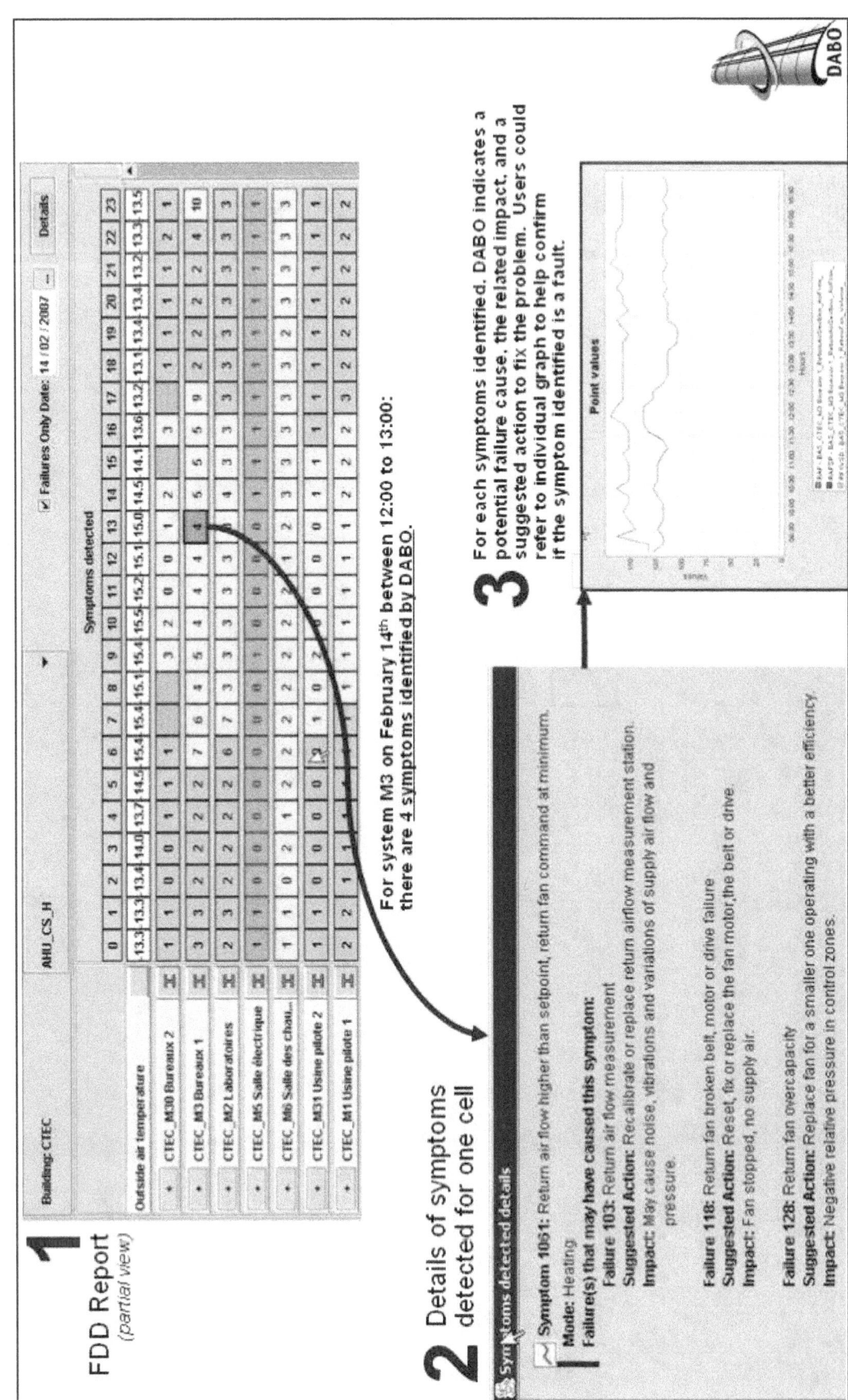

Figure 4-67 Example of Fault Detection and Diagnosis Reports

127

Point Reporter (Graph and Data)

These reports display data from building points selected by the user. Data can be shown as a line chart or a table. These reports can be requested from the FDD report, from the commissioning report, or from the "Tools" menu.

The Points Reporter displays raw data and the results of simple computations in graphic or table format for the relevant data collected or computed during the 4 hours before and 4 hours after a symptom is detected. The data displayed are sorted by date and time. Using the "Save Data" option, raw data can be extracted to a separate file for use in other applications (for example: spreadsheet software). Figure 4-68 illustrates a point reporter report.

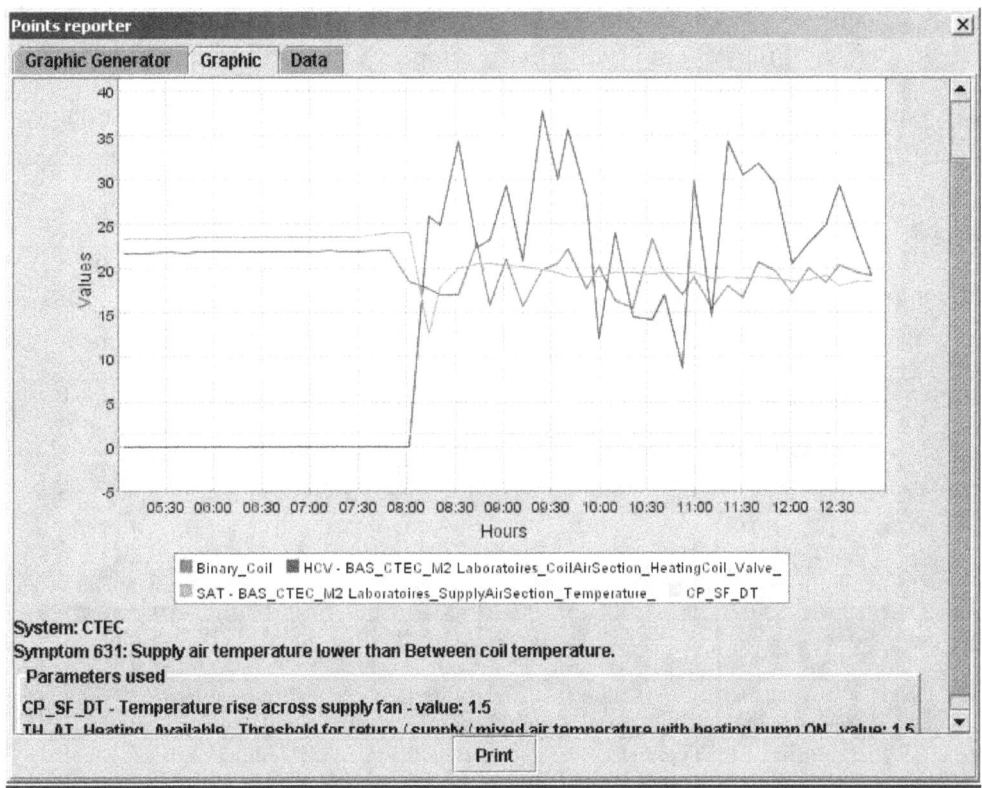

Figure 4-68. A Point Reporter Report

Commissioning Reports (COMM Reports)

These reports display data generated by the various commissioning functions or «services» or data from the building. Data are also shown in a color-coded table. These reports can be customized by the user for additional statistical analysis.

Commissioning Reports display performance values of devices on an hourly basis or summarized for an entire day, week, or month. Users can adapt the report by filtering devices by system device, system name, floor, orientation, family, and type. This allows users to observe time-based device performance variations.

This type of report is used, for example, to manually detect non-optimum set points, to identify devices with insufficient capacity, to identify systems with non-optimum sequences, to report equipment performance levels, or to report comfort performance. COMM Reports are available for all DABO™ devices.

Figure 4-69 displays performance indices for the supply temperature (PI_T) and supply temperature set point (PI_TSP) as well as the average (AVG_CZ_PI_C) and maximum (MAX_CZ_PI_C) zone connected cooling loads for two air-handling units. This report is used to detect if a temperature reset control strategy, which adjusts the supply air temperature set point as a function of the actual cooling load, performs well.

Note that the performance index for the supply temperature (PI_T) of the two air-handling units varies over the course of the day, indicating that a temperature reset strategy is being used.

For the M2 Laboratories unit, PI_T displays the same variation and has approximately the same magnitude as the average zone cooling load (AVG_CZ_PI_C). This is the expected behavior when the reset strategy is functioning correctly.

On M3, the strategy is probably not aggressive enough since while the cooling load is quite low (≈7 %), the supply air temperature index is near the mid range (≈38 %). This indicates that the supply air temperature may be too low for the given load and could generate a reheat load.

Device name	Index name	0	1	2	3	4	5	6	7	8	9	10	11	12	13	14	15	16	17	18	19	20
M2 Laboratoires	AVG_CZ_PI_C	0	0	0	0	0	0	0	0	10,3	8,8	12,5	16	20	20,5	22,2	27,2	29,2	0	0	0	
	MAX_CZ_PI_C									30,1	26,5	32,2	37,3	43,7	44,3	46,6	64,2	72				
	PI_T								-34,9	17,9	3,9	10	15,4	24,5	26,4	28,9	33,1	35,8				
	PI_TSP								15,1	-2,9	7,5	7,9	8,2	5,9	4,9	4,4	6,9	8,5				
M3 Bureaux 1	AVG_CZ_PI_C						0		0,1	13,7	11,4	10,9	5,7	6,5	6,1	7,4	5,9	3,9	0,6	0	0	
	MAX_CZ_PI_C								3	70	76,9	65,1	31,1	35,6	32,5	43,6	42,3	35,5	13			
	PI_T						-44,7	-46	-35,2	60,4	70,8	64,1	44,7	36,4	34,1	38,1	38,7	30,6	16,2			
	PI_TSP						11,9	-1,2	10,1	6,3	1	-1,1	-1	0,3	1,1	1,5	-2,3	-1,4	0,5			
M30 Bureaux 2	AVG_CZ_PI_C				0	0	0	0	0,7	11,4	18,2	22,2	21	16,8	16,3	13,2	13,9	8	0	0	0	
	MAX_CZ_PI_C								4	28	42,4	45,1	49,6	50,6	50,5	44,4	34,9	28,5				
	PI_T								1,5	16,5	33,1	38,2	46,1	49	48,7	56,1	56,2	57,6	52,9			
	PI_TSP								1,9	0,2	2	2,2	1,5	1,4	1,2	-0,9	0,8	2	-1,4			

Figure 4-69. Typical Commissioning Report for 3 Air Handling Units

Figure 4-70 shows a daily VAV box performance report. Performance indices identified by the letter 'a' consist of the VAV box basic monitored control points (supply air flow ≈ 43 l/s, supply airflow set point ≈ 45 l/s, ATD supply air temperature ≈ 18 °C and heating coil valve command ≈ 0 %). Those identified by the letter 'b' are performance indices for the air terminal device airflow ≈ 30 %, airflow set point ≈ - 1 %, cooling load ≈ 16 %, damper ≈ 43.8 % and heating load = 0 % , while those identified by the letter 'c' display information from the zone and AHU (AHU supply air temperature ≈ 20 °C, supply fan command '1= on' and zone air temperature = 23 °C).

Data are shown in grey to indicate that the AHU that supplies air to the zone is not operating during these hours. The data indicate that the VAV-reheat box is supplying 30 % of its design airflow, the damper position is approximately 43 % open, the reheat coil valve is closed, the cooling load is close to 17 %, and the heating load is at 0 %. The data also indicate the ATD supply air temperature is slightly higher than the AHU supply air temperature. This could be due to a small heating valve leak, or could indicate the need to recalibrate one or both sensors. It could also be due to normal heat gain as the air travels through the air distribution system.

Device name	Index name	3	4	5	6	7	8	9	10	11	12	13	14	15	16	17	18
1211-Photo_ATD	ATD_SAF	0	0	0	0	35	28,7	38,4	69	48	45	44,5	43,5	40,4	44,1	0	0
	ATD_SAFSP	148,8	148,8	148,8	145	104	30	38,9	68,6	47,6	45,6	45,6	45,2	39,8	43	124,8	150
a {	ATD_SAT	23	23	23	23	20,7	18,9	19,6	18,9	18,4	18,3	18,3	17,9	18,1	18,4	21,2	22,9
	AVG_ATD_HCV	0	0	0	0	0	0	0	0	0	0	0	0	0	0	0	0
	PI_ATD_AF	0	0	0	0	23,3	19,1	29,8	45,9	32	29,9	29,6	28,9	26,9	29,4	0	0
b {	PI_ATD_AFSP	-100	-100	-100	-100	-66,3	-4,2	-1,6	0,5	0,8	-1,4	-2,4	-3,8	1,6	2,6	-100	-100
	PI_ATD_C					6,2	0	13,3	22	17	15,9	16	16,4	14,6	15,5	0	0
	PI_ATD_D	98,7	98,7	98,7	98,7	73,6	38,9	42	52,7	45,7	43,6	43,4	43,6	41,9	41,7	98,7	98,7
	PI_ATD_H	0	0	0	0	0	0	0	0	0	0	0	0	0	0	0	0
	SAT	23,4	23,5	23,5	23,6	19,8	17,4	18,2	17,6	17	17	17	16,6	16,9	17,1	18,5	20,3
c {	SF1	0	0	0	0	0,5	1	1	1	1	1	1	1	1	1	0	0
	ZAT	23	23	23	22,9	22,8	22,9	23,1	23	23	23	23	23	23	23	23,3	23,4

Figure 4-70: Air Terminal Devices – typical Commissioning Report

Figure 4-71 shows the heating network performance report for one day. The performance indices are the average connected heating load (AVG_PI_H $\approx$ 22 %), maximum heating load (MAX_PI_H = 100 %), supply water pressure N_SIP $\approx$ 157 pa, supply water pressure set point N_SISP $\approx$ 140 pa, supply water temperature N_SIT $\approx$ 76 °C, supply water temperature set point N_SITSP $\approx$ 53 °C, and the temperature control valve N_V = 100 % bypass.

In this case, the heating hydronic network operates at 6 % of its full capacity (PI_DEV $\approx$ 6 %), temperature and pressure indices are over their design normal range of operation (PI_T and PI_P are over 100 %), and the temperature and pressure are significantly higher than their respective set points, making temperature and pressure set point indices very positive (PI_TSP and PI_PSP = 12 and 48 %).

Recall that the set point indices are normally approximately equal to zero when the temperature or pressure is maintained at its set point value. Since the valve command is 100 % in bypass and there is a heating load on the circuit, this report indicates that the valve and actuator should be inspected to ensure there is no leakage and/or problem with the actuator linkage.

Device name	Network type	Index name	4	5	6	7	8	9	10	11	12	13	14	15	16	17	18	19	20
boîtes Term...	Heat	AVG_PI_H	27,3	27,3	27,3	30,4	27,3	22,1	23,1	22,4	23,1	22,9	23,3	22,8	21,7	0	0	0	0,3
		CP_NET_CONV	4,2	4,2	4,2	4,2	4,2	4,2	4,2	4,2	4,2	4,2	4,2	4,2	4,2	4,2	4,2	4,2	4,2
		CP_N_SIF_HIGHLOAD	2,8	2,8	2,8	2,8	2,8	2,8	2,8	2,8	2,8	2,8	2,8	2,8	2,8	2,8	2,8	2,8	2,8
		CP_N_SIF_LOWLOAD	1	1	1	1	1	1	1	1	1	1	1	1	1	1	1	1	1
		CP_N_SIP_HIGHLOAD	150	150	150	150	150	150	150	150	150	150	150	150	150	150	150	150	150
		CP_N_SIP_LOWLOAD	130	130	130	130	130	130	130	130	130	130	130	130	130	130	130	130	130
		CP_N_SIST_HIGHLOAD	70	70	70	70	70	70	70	70	70	70	70	70	70	70	70	70	70
		CP_N_SIST_LOWLOAD	50	50	50	50	50	50	50	50	50	50	50	50	50	50	50	50	50
		CP_PUMP_NBSMAX	1	1	1	1	1	1	1	1	1	1	1	1	1	1	1	1	1
		CP_SIET_MAX	32	32	32	32	32	32	32	32	32	32	32	32	32	32	32	32	32
		MAX_PI_H	100	100	100	64,4	100	100	100	100	100	100	100	100	100				3,3
		N_SIET																	
		N_SIP	48,9	50,6	51,5	106,6	153,2	161,4	160,8	159	160,5	158,3	154,3	158,1	158	160,3	58,7	58,3	58,2
		N_SIPSP	139,6	139,6	139,6	139,6	139,6	139,6	139,6	139,6	139,6	139,6	139,6	139,6	139,6	139,6	139,6	139,6	139,6
		N_SIST	80	89,3	91,4	90,2	92	92,1	93,3	90,8	86,2	80,9	77,4	78	80,4	78,2	75	69,9	66,8
		N_SISTSP	68,3	68,4	69,4	70,2	68,1	66,5	65,4	63,4	61,8	60,5	59,1	58,5	59,3	59,2	60,4	59,4	58,4
		N_V	100	100	100	100	100	100	100	100	100	100	100	100	100	100	100	100	100
		PI_DEV	0	0	0	10	10,4	10,5	10,8	10,1	9	7,7	6,8	6,9	7,5	7	0	0	0
		PI_ET																	
		PI_F																	
		PI_FSP																	
		PI_P	-405,2	-397,1	-392,6	-117	116	156,7	153,9	144,9	152,3	141,4	121,6	140,6	140	151,4	-356,4	-356,7	-358,9
		PI_PSP	-64,9	-63,7	-63,1	-23,6	9,7	15,5	15,1	13,8	14,9	13,3	10,5	13,2	13,1	14,8	-57,9	-58,2	-58,2
		PI_PU	0	0	0	5	5	5	5	5	5	5	5	5	5	5	0	0	0
		PI_T	150	196,5	207	201	209,9	210,2	216,6	203,9	180,9	154,6	136,7	139,9	151,8	141,1	125	99,4	83,8
		PI_TSP	17,2	30,8	31,8	28,6	35	38,4	42,6	43,1	39,5	33,8	30,9	33,4	35,6	32,1	24,1	17,5	14,2
		SUM_PI_PU_S	0	0	0	5	5	5	5	5	5	5	5	5	5	5	0	0	0

Figure 4-71. NetNode Commissioning Report

Fault Management Report (FMR)

This type of report displays, by device type, confirmation of a fault repair.

The Fault Management Report (FMR) is used by the building operator/manager to manage information concerning confirmed faults in the building's operation detected by the DABO FDD module and subsequent operational corrective actions needed or done. In this report the user can review actual and historical data for a specific device, for all devices of the same type, or for all devices of a building.

Function of the information required: the FMR report can be displayed and printed in a summary or detailed format.

The following information is reviewed in FMR reports:

Device type, System name, Fault number, Fault description, Fault start time, Confirmation date, Fault priority, Primary benefit, Secondary benefit, Impact on energy savings, Cost, Responsible operator, Operator's comments, Confirmation of repaired fault and Repair technician's comments

Figure 4-72 shows for all air handling units in a building a summary display of all faults detected that need to be repaired. Figure 4-73 is an example of a detailed 'repaired faults' report. This report is useful to produce a status report of actions done during a specific period.

Graphic View	ControlZone FDD report	1221 FDD report	Fault Management Report CTEC

Family: AHU_CS_H   Device: All   Benefit: All   Timetable: Last week

Confirmed failures:

System	Failure #	Failure description	Failure start	Confirmed date	Priority	Primary benefit	Secondary benefit	Energy savings	Cost ($)	Operator	C
M3 Bureaux 1	31	Point in manual	2007-04-20 08:00	2007-05-07 13:00	Extreme	Indoor environment	Operation and mai...	Yes		Daniel Choinière	cl
M3 Bureaux 1	19	Control software problem	2007-02-19 05:00	2007-03-20 15:00	Moderate	Indoor environment	Operation and mai...	No			se
M2 Laboratoires	19	Control software problem	2007-06-19 06:00	2007-07-24 16:00	Moderate	Indoor environment	Operation and mai...	No		Daniel Choinière	se
M4 Corridor	19	Control software problem	2007-02-19 11:00	2007-03-21 12:00	Moderate	Indoor environment	Operation and mai...	No			cl
M5 Salle électrique	14	Outdoor air damper failure	2007-03-14 09:00	2007-03-21 12:00	Low	Indoor environment	Operation and mai...	Yes			ot
M5 Salle électrique	13	Mixed air damper failure	2007-06-08 00:00	2007-07-09 10:00	Low	Indoor environment	Operation and mai...	Yes		Daniel Choinière	A
M4 Corridor	13	Mixed air damper failure	2007-02-13 08:00	2007-03-21 12:00	Low	Indoor environment	Operation and mai...	Yes			cl
M2 Laboratoires	12	Exhaust air damper failure	2007-04-15 21:00	2007-05-16 19:00	Low	Operation and mai...	Indoor environment	No		Daniel Choinière	ve
M30 Bureaux 2	128	Return fan over capacity	2007-05-29 14:00	2007-06-26 16:00	Low	Operation and mai...	Asset value or tena...Yes			Daniel Choinière	cl
M2 Laboratoires	6	Return air humidity sensor	2007-02-13 07:00	2007-03-21 12:00	Very Low	Indoor environment	Asset value or tena...Yes				cl
M3 Bureaux 1	35	Supply air CO2 sensor	2007-05-03 15:00	2007-05-09 15:00	Very Low	Indoor environment	Liability reduction	No		Daniel Choinière	se
M20 Annexe 2	116	Supply fan motor or drive failu...	2007-07-30 09:00	2007-07-31 12:00	Very Low	Indoor environment	Operation and mai...	No		Daniel Choinière	ve

Figure 4-72 Typical Fault Management Report for Air Handling Unit in DABO™

133

Repaired Failures Report

from '2004-10-06' to '2009-10-06'

Building: CTEC

Generated on: Tuesday 06 October 2009

2007/09/10	Medium	Supply air flow measurement	C: Eric Le Bouthillier	Sensor failure, defective power supply
			R: Eric Le Bouthillier 2007/09/14	installation of a new 4-20mmA power supply
2008/07/21	Medium	Cooling valve failure	C: Daniel Choiniere	Check if there is a leak on the cooling valve. On july 21, the valve was at 0% and SAT was at 16°C.
			R: Leonard Serravalle 2008/09/04	Linkage disconnected with the valve. Repaired linkage

Page 3 of 9

Figure 4-73 Typical repaired failures report for an air handling unit in DABO™

Building case study: Ongoing Commissioning Project

The demonstration building is the CANMET Energy Building in Varennes, Québec, Canada. Built in 1992, the single-floor, 3600 m^2 building includes office space for 90 people, two laboratories, two industrial pilot plants, conference rooms and a cafeteria.

The building, designed to be energy efficient, incorporates low energy technologies such as a passive solar preheating device, ice bank storage, photovoltaic cells, as well as a central gas heating plant and a central electric chilled water plant. Each area of the building is served by a specific air system designed for its occupation. The HVAC systems are centrally controlled by a BEMS system. (Table 4-18)

Table 4-18 CTEC-Varennes HVAC Systems

HVAC systems	Capacity	Location
Heating		
Fire tube boilers (2)	470 kW each	Building
1 primary and 5 secondary hydronic circuits, 7 pumps, constant volume		
Cooling		
1 air cooled chiller	406 kW	Building
2 ice bank tanks	1145 kW-h	Building
1 hydronic circuit, 2 pumps, constant volume		
Air Handling system		
M1 (CAV, HEA)	2, 735 L/s	Pilot plant1
M2 (VAV, 100 % fresh air, HEA,CO)	5,815 L/s	Laboratories
M3 (VAV, HEA, CO)	5,500 L/s	Office phase 1
M4 (CAV, HEA)	1,265 L/s	Storage phase 1
M5 (CAV)	160 L/s	Mechanical room
M6 (CAV)	1,030 L/s	Boiler room
M30 (VAV, HEA, CO)	1,660 L/s	Office phase 2
M31 (CAV, HEA)	5,200 L/s	Pilot plant 2
M32(CAV)	2,000 L/s	Mechanical room 2

The ongoing commissioning process started in 1999 and continued until 2009. It aimed at resolving operating problems, improving comfort, optimizing energy use and recommending retrofits where necessary. Delivery of the ongoing commissioning project system included a series of tasks performed in four steps: planning, investigation, implementation and hand off (see Table 4-19). Tasks surveyed with DABO are shown in italics. As it is an ongoing commissioning process, the investigation and implementation have been gradually and continuously performed over the 2000 to 2006 period. Since 2006, DABO is still used on a regular basis to ensure the persistence of savings and detect new deficiencies.

Results for the investigation and implementation period 2000 to 2006 were presented in 'Four Years of Ongoing Commissioning in CTEC-Varennes Building with a BEMS Assisted Cx Tool'[44] (Choinière 2004) and are summarized in the following section.

Results for the hand off and persistence period include new deficiencies that occurred during normal operation, as well as deficiencies that were not detected during commissioning of the various optimization projects (2006 to 2009).

Table 4-19   Ongoing Commissioning Project Delivery System at CETC-V

PLANNING
• Choose team
• Define project objectives, scope and deliverables
• Review building documentation and energy bills
• Develop commissioning plan
• Initiate cooperation with building operation team
INVESTIGATION (continuous over 6 years, 2000-06)
• Assessment
○ Site, design and occupant needs assessment
• Installation of DABO
• Develop and carry out diagnostic tests and system monitoring
• Analyze monitoring results
• Develop list of deficiencies and improvements
○ Include capital improvement opportunities
○ Include training recommendations
• Select the most cost effective opportunities
IMPLEMENTATION (continuous over 6 years, 2000-06)
• Implement improvements identified in investigation phase
• Retest and re-monitor to confirm the results
• Adjust, if necessary, improvements carried out during investigation phase
• Review energy consumption reduction estimates
• Building Operator training and occupant information
HAND OFF- PERSISTENCE (continuous since 2006)
• Prepare and present final report
○ As-Built Re-Commissioning work
○ New sequence of operations manual
○ Testing and balancing (TAB) report (air, water)
○ Energy baseline
○ Check-up of energy bills (3 months)
○ Proposal for EE measures with longer payback
• Implement an ongoing commissioning process and energy management plan
○ Ensure that the use of DABO is well understood by operators so as to maintain re-commissioning benefits

Results for Investigation and Implementation Step (1999 to 2006)

- Implement a continuous energy management plan 1998 (in house staff)

- Reset operation schedules (AHU, hydronic circuits) (1999)

- Optimize controls and sequence of operation

  o Function of actual needs

  o Peak load management (chiller, humidification) (2003)

  o Avoid simultaneous heating and cooling

- Reset set points (AHU, hydronic circuits) (2000 to 2005)

  o Minimum fresh air

  o Supply pressure and temperature

  o Night set back

- Fixed minor deficiencies

  o Sensor calibration

  o Low heating capacity in some rooms

  o Replacement of leaking valves

- Investment in measures with short payback

  o Addition of DDC controls (chiller, boiler 2001)

  o Link AHU M2 to solar wall (March 2003)

  o VSD on 3 fans (March 2002)

- Energy efficiency project (2005 to 2006) $250,000

  o Conversion of pneumatic room controls to DDC

  o Off peak electric boiler (200kW)

  o AHU M2 100 % fresh air convert to a recirculation system

  o Connection on AHU M4 fresh air to solar wall

  o VSD on 6 pumps and fan motors

- Energy reduction

  o Figure 4-74 shows the impact on the energy consumption of the ongoing commissioning project implemented at the CETC-V since 1998. During this period 1999 to 2006, measures implemented have resulted in a 35 % reduction in electricity usage and 45 % reduction in natural gas consumption.

Results for the Hand-off and Persistence Phases (2006 to 2008)

Table 4-20 presents results, in terms of the number of faults detected but not yet repaired, summarized and classified as a function of the type of fault, the level of priority to repair the fault and the impact on the energy used. Table 4-21 presents results, in terms of number of faults that were detected and repaired, summarized and classified as a function of the type of fault, the level of priority to repair the fault and the impact on the energy used.

Types of faults include:

- Sensor fault: This includes complete failure and incorrect reading of sensors

- Defective output: This includes stuck, incorrect feedback position, and defective leakage of valve, damper, fan, humidifier, coil, controller, fan and pump motor, etc.

- Control program fault: logic (set point and output, tuning, signal instability), inappropriate sequence of operation, incorrect minimum position, non optimal control set point and schedule, control point set in manual

- Installation problem: inappropriate control device installation or connection to control system, undersized or oversized components,

Priority to be repaired includes:

- None: These faults have little impact on energy consumption or occupant comfort and are difficult to repair without significant modification to the system design.

- Low: These faults usually have no impact on comfort and energy. They only affect the information provided to a user in the diagnosis and analysis of control system equipment.

- Moderate: These faults have an impact on energy consumption or the behavior of all system operations. The can also have an influence on the expected life duration of equipment problems.

- High: These faults usually have considerable impact on occupant comfort or a high impact on energy consumption

Impact on the energy consumed: This criteria indicates if the fault causes an increase in energy use.

Number of faults detected but not repaired during 2006 to 2009					
Type of fault	Priority				Energy
	None	Low	Moderate	High	
Sensor		8	2		1
Defective output		1			1
Control logic	2	9	2		2
Installation	5	2	1		2
Total	7	20	5		6

During 2006 to 2009, a total of 126 faults occurred in the building energy system; 94 have been repaired while 32 are still in effect. Those repaired had no impact on energy consumption or comfort; no major changes to system design were required. In terms of categories, 35 relate to sensors, 37 to output, 39 to logic fault while 15 stem from problems related to installation or design. A total of 54 faults have or had an impact on energy consumption. A total of 116 faults were automatically detected by the various DABO™ Fault detection and Diagnosis services while 10 were detected manually or analyzed using DABO™ Commissioning Reports.

Table 4-21. Number of Faults Detected and Repaired during 2006 to 2009

Number of faults detected and repaired during 2006 to 2009					
Type of fault	Priority				Energy
	None	Low	Moderate	High	
Sensor		20	4	1	12
Defective output		31	3	2	20
Control logic		20	3	3	12
Installation	3	2	1	1	5
Total	3	73	11	7	49

Energy reduction

Figure 4-74 shows the impact on energy consumption of the ongoing commissioning project implemented at the CETC-V since 1998.

During this period, measures implemented resulted in a 43 % reduction in electricity usage and an 81 % reduction in natural gas consumption. In 2008 to 2009, cost savings were $91,861 CDN, representing 49 % of the building's energy bills. Cumulative savings since 1998 are $592,281 CDN while the whole-building energy use dropped from 2248 MJ/m² to 1016 MJ/m².

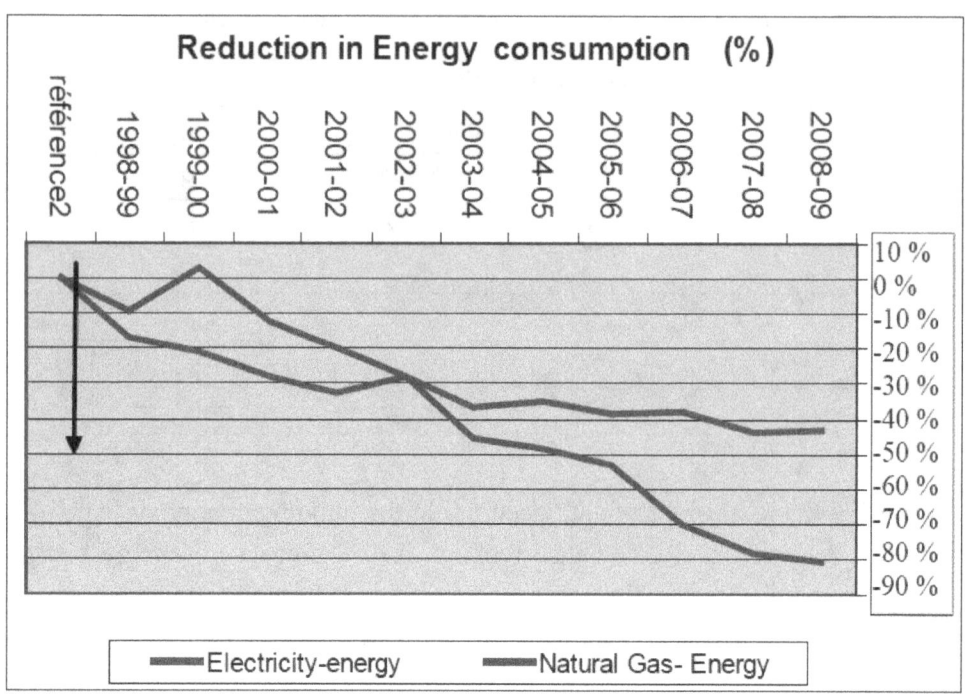

Figure 4-74 . Ongoing Commissioning Impact on CETC-V Building Energy Consumption

Conclusions

An ongoing commissioning process ensures that buildings reach and operate at their optimized energy cost and performance levels, while ensuring comfortable conditions for occupants. The CANMET Energy project generated a 43 % reduction in the use of electricity and a 81 % reduction in natural gas consumption during 1998 to 2008.

DABO, a BEMS-assisted commissioning tool, monitored the enormous amounts of data produced and provided an extensive analysis of incoming data.

The use of a BEMS-assisted commissioning tool has helped to circumvent commissioning barriers by automating parts of the process, which has reduced the costs for commissioning. Developing a detailed, systematic and automated approach has improved the quality assurance process and overall performance of the building. Furthermore, automating this essentially manual process has allowed its application on an ongoing basis, generating benefits over the entire life of the CETC-Varennes building.

The optimization process and the 'DABO^{TM}' tool are being demonstrated in more than 10 projects. Demonstration projects include some of the first Canadian LEED buildings and the participation of major Canadian facility management firms and commissioning providers.

DABO^{TM} is a tool in constant evolution. Current research efforts are concerned with the development of an energy predictor, new fault detection and diagnosis modules for heating and cooling networks, and new analyses of BEMS data to enhance the commissioning process.

4.4.6 Optimization of Automation Functions with Dynamic Simulation as the Energy Management System (OASE)

Office buildings and commercial buildings, particularly those subject to stringent requirements regarding comfort and usage, comprise highly complex systems. They are designed using an integrated approach and are marketed as "innovative" or "intelligent". It is crucial that these buildings be fine-tuned and optimized after their initial startup. Investigations show that, during operation, many buildings do not meet their target energy and thermal comfort performance; this could be due to the fact that the time allocated for commissioning is often too short. Additionally, ongoing optimization is rare. More specifically, energy-efficiency-based adjustments to accommodate changes in building systems' equipment or building occupancy are often not made and the influence of parallel system operation is often ignored.

The right course has to be set out during design; the optimization of building systems cannot be left until after operation has begun. The design goal is sustainability throughout the whole lifecycle of buildings and their systems. Requirements for energy efficiency and occupant comfort must be considered and implementable at planning, system design and energy management stages. Building optimization must be made possible during operation.

What methods and tools are needed? Answers were found in the research project "OASE" (Optimization of Automation Functions with Dynamic Simulation as the Energy Management System), funded by the German Federal Ministry of Economics and Technology (BMWi). With the methodology of Operation Diagnostics developed in this R&D project, recorded building automation data is used to identify weak points in terms of energy performance, as well as any operational shortcomings, and to derive optimization potential from them. The OASE methodology has since been tested in a variety of demonstration buildings. It supplements current research activities with the goal of advancing the introduction of holistic building energy concepts in practice. This includes the development of "cost-effective energy management for residential and office buildings" (KENWO – "Kostengünstiges Energiemanagement für Wohn- und Bürogebäude") and development of an "Energy Navigator," which enables partially automated enhanced building operations by means of interfaces with the building automation system.

OASE – Methods and Procedures

It is worthwhile to evaluate office buildings and commercial buildings over several years, examining and assessing their energy consumption and occupant comfort based on measurement. With this strategy, laborious analysis methods and costly measurement equipment can be avoided. The Operation Diagnostics process developed in the research project OASE provides a standardized methodology and various tools that use and analyze the available energy-related data from the building automation systems. This makes it possible to identify shortcomings and malfunctions in building operation and to derive optimization measures. The results are improved building performance and effective building management.

Measurement Data Visualization and Operation Diagnostics

In OASE, Operation Diagnostics, the visualization of measurement data aids in the detection of symptoms of malfunction and identifies optimization opportunities, even in complex systems. Measured data points from building automation or energy management systems are graphically displayed. In so-called "carpet plots", color-coded measurement values are shown in bars (one bar for each day), which are arranged sequentially over the measurement period (see Figure 4-78). Operating states that recur in daily and weekly cycles become identifiable as patterns. In scatter plots, the same data is plotted so as to display the interdependencies between various control parameters. Thus, structures and patterns, or deviations, can be identified quickly, and conclusions made regarding the operation of the building's mechanical systems. With Operation Diagnostics, the systems' functionality and performance can be examined and assessed. The goal is to optimize control parameters and detect flawed, ineffective or inefficient system and building operation.

Operation Prognostics

Operation Prognostics is a methodology to incorporate dynamic building operation into design. Operating functions and control sequences are developed in the early design phases, optimized for the building and the technology used and documented in the form of operation patterns, the same graphical means used for the Operation Diagnostics. During the design phase, the operational concept is repeatedly adjusted and refined. Simulation programs can be used for optimization. Operation Prognostics thus allows better use and implementation of the design solution during operation. The goal is to save as much energy as possible during operation, minimizing the building's lifecycle costs and environmental impact.

Holistic Building Operation

System optimization should not be left until after operation has begun; it must be taken into consideration during design. This applies to the sizing, as well as the interaction of separate systems and system components. Thus, in the course of Operation Prognostics, methods and tools should be developed that facilitate the integration of operational strategies into planning and operation. This is to establish the link between integrated building design and holistic facility management.

Early Applications of Operation Diagnostics

Operation Diagnostics systematically analyzes key energy-related data from the already installed building control system in its current state and enables it to be used for optimization of operations involving heating, cooling, ventilating, air-conditioning and lighting. It focuses on measures that require no investment and that can be implemented with existing system technology. In order to verify its benefits to architects, owners, mechanical system designers and building operators, as well as to test its applicability to various system concepts and building types, the methodology was put through field tests in several demonstration buildings.

Building case study 1: The Munich Re East Building

This office building was built in 1965 as a reinforced concrete structure with an aluminum curtain wall facade. The mechanical system was replaced in 1997. The building is partially air-conditioned and ventilated. Centralized cooling is provided by means of a chilled water system and free cooling; heating is provided by means of a district heating connection. In the OASE project, Munich Re's east building was subjected to operation diagnostics combined with a measurement program to record additional temperature values and flow rates in the heating and cooling loops. From the analysis of the operation data, a series of quick-win measures for the heating and cooling systems were derived, which enables low to no-cost optimization of operational management. For instance, it was shown that energy savings could be achieved by reducing pump run times and optimizing control of the cooling loops. The data analysis enabled comparison between the set operating times and the usage requirements, as well as verification of sensors' functionality. In addition, the project provided a valuable base test for the visualization and analysis procedures of operation diagnostics and for optimizing the methodologies, tools and procedures.

Figure 4-75 The Munich Re East building, Königinstrasse in
Munich, Germany

Building case study 2: Laboratory building 'Forschungszentrum Jülich'

This 3-story laboratory from the 1960s was completely refurbished by 2002; emphasis was placed on a new concept for ventilation and cooling. The operation diagnostics was limited to individual data point visualization via carpet plots and accompanying annotation. Nevertheless, it was possible to validate the refurbishment's forecast savings potential, and the operation diagnostics helped to reveal further optimization potential regarding dynamic building and system operation. Visualization and analysis of hourly temperature data over the course of one year provided the operator with important tips for optimizing the building's supply air

145

temperature settings. The tool also enabled clear identification and display of data transfer errors. In view of an upcoming general overhaul of the operational control system at Forschungszentrum Jülich, it would be practical to incorporate an operation diagnostics tool into the daily operational control of the building.

Figure 4-76 The laboratory building before (top) and after (bottom) refurbishment

Building case study 3: Gebhard Müller School in Biberach

This new building was built in summer 2004. It was designed as a vocational school building, a so-called 3-liter house. Comprehensive building operation simulations were conducted in the design phase. As part of the research program "Energy-Optimized Building" (EnOB), the concept was measured and optimized during operation from the beginning. This was supported by operation diagnostics, resulting in numerous adjustments to building automation including the optimization of operating times.

Ground water serves as the building's main heat source and only cold source; it is used with a heat pump system in heating mode, and directly through a heat exchanger in cooling mode. With the high electricity consumption caused by the provision of ground water, Operation Diagnostics made it possible to adapt total ground water requirements to suit actual usage profile, i.e., requirements were reduced and the efficiency of the two ground water heat pumps increased. Analysis of requirements and optimization also resulted in a realignment of operation sequences. The wood pellet boiler installed solely to cover peak loads was also used to supply heat to the ventilation system at a higher temperature during the day. This operational change increased utilization of the boiler's capacity. Operation could be optimally coordinated with the concrete core activation (radiant heating and cooling), thereby increasing heat pump efficiency.

Figure 4-77. Gebhard Müller School in Biberach, Germany

Heat Pump Return Temperature

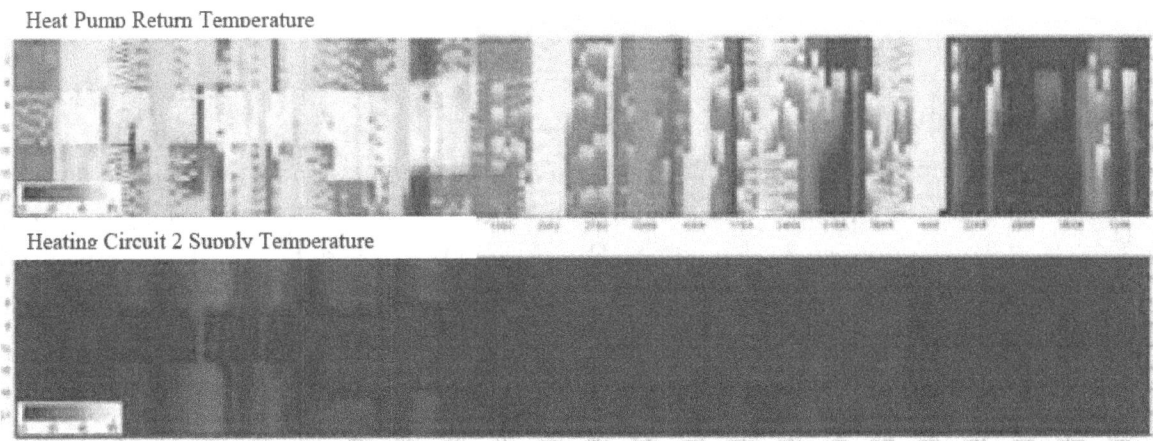

Heating Circuit 2 Supply Temperature

Figure 4-78. Carpet plot of heat pump operation in the Gebhard-Müller School (excerpt)

Conclusion

Tools developed in the OASE project offer new ways to visualize measurement data and thus facilitate a systematic, goal-driven analysis of building operation. Their practical benefits for architects, mechanical system designers, building operators and owners have been demonstrated through several actual-building applications. Currently, basic functions of the OASE tool can be found in a number of building automation systems. The detailed analysis options are also gradually being introduced into ongoing optimization processes.

It remains to be seen whether more widespread application of operational diagnostics will affect building design practice. Ultimately, for energy-saving building operation, the decisive course has to be set during design rather than after operation. In practice, however, obstacles are often met in the implementation of identified savings opportunities, which cannot be overcome with analysis tools alone.

For instance, in buildings with facility management, staffs are often unfamiliar with systems, usage periods and comfort requirements. Furthermore, service contracts usually contain general statements on energy optimization, which are not linked to remuneration provisions. No motivation or incentive is provided for the identification and implementation of performance measures. The service provider benefits from energy savings only after a while – if at all.

Operation Diagnostics contribute significantly to energy-efficient building operation. It "refines" measurement values and data from building controls and automation, producing findings that enable systematic optimization measures. Deficiencies, malfunctions and system defects can be identified sooner rather than later, even if the building occupancy changes. As a result, motivated building occupants can be proactive, and the building operator can react before complaints arise without having to rely on an expert witnessing the system malfunction or defect at the time that it occurs.

References Chapter 4

[20]   IEA. 2004. International Energy Agency. Annex 40: Commissioning of Buildings and HVAC Systems for Improved Energy Performance. http://www.commissioning-hvac.org.

[21]   The Energy Performance of Buildings, Directive 2002/91/EC of the European Parliament and of the Council, European Union, 2002.

[22]   M. Frank et al. "State-of-the-Art Review for Commissioning Low Energy Buildings: Existing Cost/Benefit and Persistence Methodologies and Data, State of Development of Automated Tools and Assessment of Needs for Commissioning ZEB" NIST Interagency/Internal Report (NISTIR) 7356. 2007.

[23]   J. Hyvarinen, S. Karki (Eds.), Building Optimization and Fault Diagnosis Source Book, IEA Annex 25, VTT, Finland, 1996.

[24]   W.Z. Haung, M. Zaheeruddin, S.H. Cho, Dynamic simulation of energy management control functions for HVAC systems in building, Energy Conversion and Management 47 (2006) 926–943.

[25]   L. Lu, W. Cai, L. Xie, S. Li, Y.C. Soh, HVAC system optimization – in building section, Energy and Buildings 37 (2005) 11-22.

[26]   K.F. Fong, V.I. Hanby, T.T. Chow, HVAC system optimization for energy management by evolutionary programming, Energy and Buildings 38 (2006) 220-231.

[27]   Z. Ma, S.W. Wang, X. Xu, F. Xiao, A supervisory control strategy for building cooling water systems for practical and real time applications, Energy Conversion and Management 49 (2008) 2324-2336.

[28]   T.I. Salsbury, R.C. Diamond, Fault detection in HVAC systems using model-based feedforward control, Energy and Buildings 33 (2001) 403-415.

[29]   P. Haves, R.J. Hitchcock, K.L. Gillespie, M. Brook, C. Shockman, J.J. Deringer, K.L. Kinney, Development of a Model Specification for Performance Monitoring Systems for Commercial Buildings, Proceedings of the 2006 ACEEE Summer Study on Energy

Efficiency in Buildings, Pacific Grove, California, Lawrence Berkeley National Laboratory, 2006.

[30] M.A. Piette, S.K. Kinney, P. Haves, Analysis of an information monitoring and diagnostic system to improve building operation, Energy and Buildings 33 (2001) 783-791.

[31] S.W. Wang, F. Xiao, AHU sensor fault diagnosis using principal component analysis method, Energy and Buildings 36 (2004) 147–160.

[32] J. Liang, R. Du, Model-based Fault Detection and Diagnosis of HVAC systems using Support Vector Machine method, International Journal of Refrigeration (2007) 1-11.

[33] S.H. Cho, H.C. Yang, M. Zaheer-uddin, Byung-Cheon Ahn, Transient pattern analysis for fault detection and diagnosis of HVAC systems, Energy management and conservation 46 (2005) 3103–3116.

[34] J. Schein, S.T. Bushby, N.S. Castro, J.M. House, A rule-based fault detection method for air handling units, Energy and Buildings 38 (2006) 1485–1492.

[35] S. Katipamula and M. Brambley, 2005 Methods for Fault Detection, Diagnostics, and Prognostics for Building Systems—A Review, Part I, HVAC&R Research Volume 11, Number 1.

[36] S. Katipamula and M. Brambley, 2005 Methods for Fault Detection, Diagnostics, and Prognostics for Building Systems—A Review, Part II, HVAC&R Research Volume 11, Number 2.

[37] P. Haves , D. Claridge, M. Lui, Report assessing the limitations of EnergPlus and SEAP with options for overcoming those limitations, California Energy Commission Public Interest energy Research Program, HPCBS #E5P2.3T1, 2001.

[38] F. Xiao, S.W. Wang, Progress and methodology of lifecycle commissioning of HVAC systems to enhance building sustainability, Renewable and Sustainable Energy Reviews (In Press) (2008).

[39] "Sustainable control and maintenance in buildings", Quick scan protocol, TNO, January 2008.

[40] Jean Gilles. 2004. A climate change solution; "Intelligent Building Operating Technologies". http://cetcvarennes.nrcan.gc.ca.

[41] Choinière D. 2001. Un agent de détection et diagnostic de fautes pour les bâtiments. Congrès de l'Association québécoise pour la maîtrise de l'énergie. Quebec City, Quebec, Canada.

[42] IEA. 2001. International Energy Agency. Annex 34: Computer-Aided Evaluation of HVAC System Performance. Final Report. Editors Arthur Dexter and Jouko Pakanen.

[43] D.Choinière, and M Corsi. 2003. A BEMS-assisted commissioning tool to improve the energy performance of HVAC systems. Proceedings of ICEBO 2003. Berkeley, CA.

[44] D.Choinière. 2004. Four years of Ongoing commissioning in CTEC-Varennes Building with a BEMS-Assisted Cx Tool. Proceedings of ICEBO 2004. Paris, France.

5   Conclusions

Commissioning – and especially ongoing commissioning – must cope with vast amounts of information and data. Information on the building's characteristics and HVAC system, as well as measured data, e.g., of the energy consumption, state variables or control signals, must be managed. In more complex or low energy buildings with advanced technologies this task requires computer-aided processes or software tools because humans cannot manage the wealth of data or extract specific information.

The tools and case studies discussed illustrate the potential for improved building performance through the use of software tools in the commissioning process. The tools provide features for operational fault detection and diagnosis (FDD) and/or for optimization that help to identify and realize potential savings.

Most developed tools focus on the operation phase (existing buildings). Only a handful address the design phase (new constructions). This seems natural as commissioning has close ties to the operation phase and most of the built environment already exists. However, ideally, commissioning starts in the early design phase of construction. In this context, it should be noted that design and operation phases demand very different tool features:

Tools for design phase (new construction)

The design phase is characterized by considerable choice in solutions for building construction and HVAC systems. The design team must find a way to meet the owner's project requirements (OPR). Typically this involves comparing various design alternatives that (ideally) are based on simulation. Consequently, a design phase Cx tool should support the assessment of different design alternatives.

More important is documentation of the OPR, basic design, target values (e.g., energy demand), and all decisions and knowledge gathered during the design phase. This information is essential to measurement and verification in the latter operation phase. The current absence of complete and/or up-to-date building documentation is one of the biggest barriers to the introduction of commissioning in existing buildings. Cx tools should therefore support information management over the building's lifecycle.

Tools for operation phase (existing buildings)

In the operation phase, most commissioning is made up of FDD and optimization. Unlike in the design phase, the commissioning provider must deal with a given system without a chance to replace major equipment (as long as no significant refurbishment is planned). In fact, many existing buildings were not commissioned during the design phase and many may not even have been commissioned during hand over. As a result, the commissioning provider must often cope with a situation in which a lack of information and little measured data exist.

Consequently, the features and scope of the operation phase vary significantly, depending on the data and information needed to apply the tool.

For Example, the application of a simulation model for optimization requires detailed information about the building's structure and HVAC design. If these are unavailable (and there is insufficient budget to gather them) it cannot be used for commissioning. However, measured data may be available that can be used for visualization, e.g to create energy signatures that reveal characteristics of the energy consumption. The kind of tool that might be applied for commissioning depends on an existing building's history and documentation.

Chapter 4.3 gives an overview of developed tools.

Aside from the kind of tool, data visualization – either for manual FDD or presentation of analysis results - was identified as a vital feature by the Annex 47 team. Chapter 4.2 gives an overview. Automation is still uncommon in the commissioning process (e.g., for FDD). Visualization therefore provides experts with a valuable tool for revealing information hidden in data.

The international status of ongoing commissioning (ongoing Cx) and Cx tool requirements can be summarized as follows:

- There is no common understanding of the term "ongoing commissioning" among the different countries.
- With the exception of the USA, ongoing Cx has yet to be established as a well defined third-party service. Nevertheless, every country offers services that form part of the ongoing Cx process.
- All experts stress the importance of ongoing Cx for the persistence of energy efficient operation of buildings.
- All countries (including the US) state that there are few if any commercial and/or easy to use tools available that can be used for ongoing Cx.
- Moreover, all countries state that automation of the tools and thereby a reduction of labor cost connected to ongoing Cx is crucial for a wider application
- Cost-benefits of ongoing Cx-tool application have yet to be documented.

The Annex 47 team identified the following needs for further research and tool development to support ongoing commissioning in new and existing buildings:

- Automation and a more robust application of tools (e.g., for FDD or functional testing) remains an important issue to reduce costs.
- Tools should be better integrated, i.e., not just provide one feature like FDD for an AHU. This will make the application of tools increasingly relevant to building owners or operations staff. The number of tools required to cover all ongoing Cx-required functionalities should be reduced.
- Tools must be made easy to use and their interfaces improved.

Finally, tools should be better integrated in the whole commissioning process. To do so, the process itself must be better defined and standardized. An example is the need for monitoring guidelines discussed in Chapter 3.2.

6   Annex 47 Participants

Name	Country	Affiliation
Alexis Versele	Belgium	KaHo St-Lieven
Hilde Breesch	Belgium	KaHo St-Lieven
Stephane Bertagnolio	Belgium	University of Liege
Daniel Choinière	Canada	Natural Resources Canada
Zhongxiian Gu	China	TNO Beijing
Yongning Zhang	China	Tsinghua University
Karel Kabele	Czech Republic	Czech Technical University
Pavla Dvorakova	Czech Republic	Czech Technical University
Michal Kabrhel	Czech Republic	Czech Technical University
Mika Violle	Finland	Helsinki University of Technology
Jorma Pietilainen	Finland	Technical Research Centre of Finland (VTT)
Hannu Keranen	Finland	Helsinki University of Technology
Lari Eskola	Finland	Helsinki University of Technology
Satu Paiho	Finland	Technical Research Centre of Finland (VTT)
Hossein Vaezi-Nejad	France	Dalkia
Oliver Baumann	Germany	Ebert & Baumann Consulting Engineers
Steffen Plesser	Germany	Institute of Building Services and Energy Design (IGS)
Christian Neumann	Germany	Fraunhofer Institute for Solar Energy Systems
Dirk Jacob	Germany	Fraunhofer Institute for Solar Energy Systems
Anatoli Hein	Germany	Institute of Building Services and Energy Design (IGS)
Michele Liziero	Germany/Italy	Politecnico di Milano, Guest Scientist ISE
Jochen Schaefer	Germany	Ebert & Baumann Consulting Engineers
Shengwei Wang	HK/China	Hong Kong Polytechnic University
Xinhua Xu	HK/China	Hong Kong Polytechnic University
Zhenjun Ma	HK/China	Hong Kong Polytechnic University
Zhou Cxiang	HK/China	Hong Kong Polytechnic University
Xiao Fu Linda	HK/China	Hong Kong Polytechnic University
Na Zhu	Hong Kong	Hong Kong Polytechnic University
Zoltan Magyar	Hungary	University of Pecs
Csaba Fodor	Hungary	University of Pecs

Name	Country	Affiliation
Harunori Yoshida	Japan	Kyoto University
Motoi Yamaha	Japan	Chubu University
Mingjie Zheng	Japan	Sanyo Air Conditioning
Yasunori Akashi	Japan	Kyushu University
Hiroo Sakai	Japan	Hitachi Plant Technologies
Katuhiro Kamitani	Japan	Tonets Corporation
Ryota Kuzuki	Japan	Tokyo Gas Co
Katsuhiko Shibata	Japan	Takasago Thermal Eng. Co
Fulin Wang	Japan	Kyoto University
Masato Miyata	Japan	Kyoto University (student)
Hirotake Shingu	Japan	Kyoto University (student)
Hiromasa Yamaguchi	Japan	Kansai Electric Power Co
Ryusi Yanagihara	Japan	Tokyo Electric Power Co
Hideki Yuzawa	Japan	Nikken Sekkei Research Institute
Takao Odajima	Japan	Takenaka Corp.
Hirobumi Ueda	Japan	Osaka Gas CO., Ltd
Masahiro Shinozaki	Japan	Kyushu Electric Power Co.
Katsuhiro Kamitani	Japan	Tonets Corporation
Mingjie zheng	Japan	SANKO AIR CONDITIONING CO.,LTD
Katsuhiko Shibata	Japan	Takasago Thermal Engineering Co..Ltd
Vojislav Novakovic	Norway	Norwegian University of Science and Technology (NTNU)
Natasa Djuric	Norway	NTNU
Marko Masic	Norway	NTNU
Vojislav Novakovic	Norway	NTNU
Henk Peitsman	the Netherlands	Netherlands Organization for Applied Scientific Research (TNO)
Luc Soethout	the Netherlands	TNO
Ipek Gursel	the Netherlands	University of Delft
Natascha Milesi Ferretti	USA	National Institute of Standards & Technology
David Claridge	USA	Texas A&M University
Hannah Friedman	USA	Portland Energy Conservation Inc.

Name	Country	Affiliation
Omer Akin	USA	Carnegie Mellon University
Ashish Singhal	USA	Johnson Controls
Tudy Haasl	USA	Portland Energy Conservation Inc
Phil Haves	USA	Lawrence Berkley National Laboratory

APPENDICES

6.1 APPENDIX 1: Survey on Sensors for Energy Management in Existing Building

Three problems exist: cost, space, and owner agreement to install sensors in an existing building. We assessed the current status of these problems using the "Questionnaire on sensors for energy management in existing buildings". The questionnaire has four parts;

1. Do existing buildings have sufficient sensors to carry out energy management?
2. Do existing buildings have sufficient space for the installation of additional sensors?
3. Are affordable sensors in the marketplace equipped with sufficient accuracy for purposes of energy management?
4. Do owners demonstrate an understanding of the need for additional sensors and allow for this in their budgets?

Number of respondents:
Asia: Japan 60
North America: USA 6, Canada 1
Europe: Germany 1, the Netherlands 1, Belgium 1, Norway 3

Table 0-1 Response rates

	Japan	North America	Europe
Design offices	15 %	0 %	0 %
General contractors	23 %	17 %	0 %
Construction companies	38 %	0 %	20 %
Consultants	5 %	33 %	60 %
Others	20 %	50 %	20 %

1. Do existing buildings have sufficient sensors for energy management?

1) Survey about actual and required energy management level of existing building

We investigated actual and required energy management levels of existing buildings, using the "Building Energy Management Levels" discussed in SHASE.

【Building Energy Management Levels】

A. A total amount of consumed energy is measured for the whole building (for every energy type, monthly data).

157

B. A total amount of consumed energy is measured for the whole building (for every energy type, daily and hourly data).

C. Each energy amount categorized for HVAC, lighting, consents, elevators, etc., is measured.

D. Each energy amount categorized for office areas, conference areas, communication machine rooms, parking lots, etc.

E. Each energy amount for each floor.

F. Energy consumed by each equipment type, such as chillers, cooling towers, pumps, etc. or each sub-system, such as power plant, air-conditioning system, pump system, etc.

Figure 0-1 shows actual energy management levels. Figure 0-2 shows required energy management levels from respondents' answers.

In Japan, most buildings measure the total amount of energy consumed for the whole building, for every energy type, with monthly data (Level A). However, the percentage of Level B is less than 50 %, and that of Levels C-F less than 30 %.

Level A ratios for America and Europe are similar to those of Japan. American and European ratios of Levels B-F are less than those of Japan.

In contrast to actual energy management levels, respondents consider that each energy amount categorized should be measured: HVAC, lighting, consents, elevators, etc. (Level C), or total amount of daily and hourly energy consumed for the whole building (Level B).

Respondents in Japan answered that the required level is Level C; respondents in America answered Level B and respondents in Europe answered Levels B or C.

Many respondents who answered Level C consider that, to improve energy performance it is essential to analyze each energy amount categorized by use (like HVAC, lighting, consents, elevators, etc.). Many respondents who answered Level B consider that, from a cost benefit aspect, it would be best to measure only a few sensors but use an hourly high time resolution for greater understanding of consumption trends.

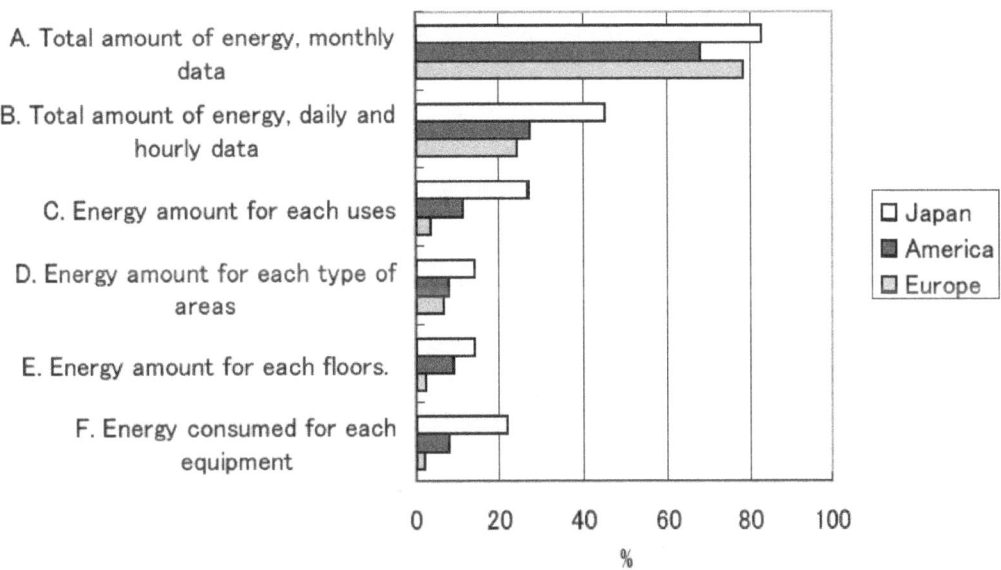

Figure 0-1 Actual energy management levels

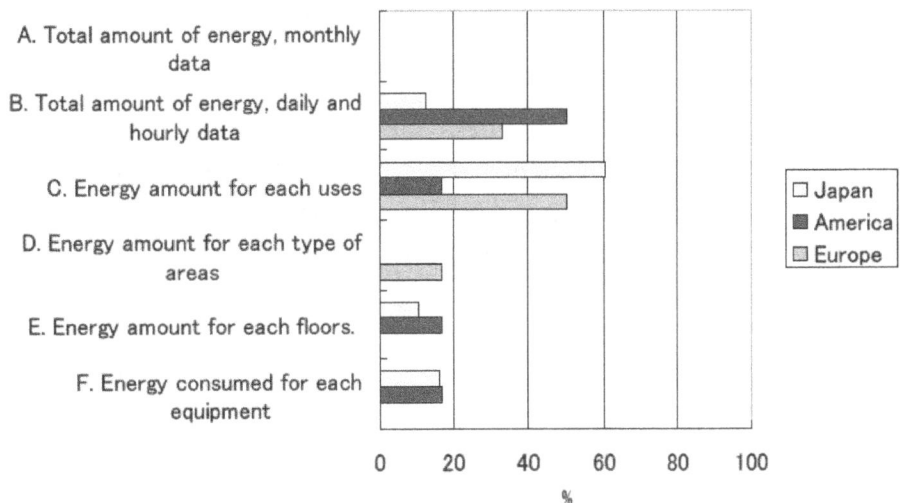

Figure 0-2 Required energy management levels

2) Survey on dissatisfaction with sensors and monitoring systems in existing buildings

We investigated points of dissatisfaction related to sensors and monitoring systems in existing buildings. Figure 0-3 illustrates the survey results. Most respondents are dissatisfied with fewer sensors in existing building. And many are dissatisfied with sensors not being calibrated. Many respondents consider sensor calibration to be important. The ratios for America and Europe are similar to those of Japan.

Overall, fewer respondents in Japan are dissatisfied with inadequate sensor accuracy, poor accuracy in monitoring systems, or miss sensor selection when compared with. respondents in America and Europe (30 % and 70 % respectively).

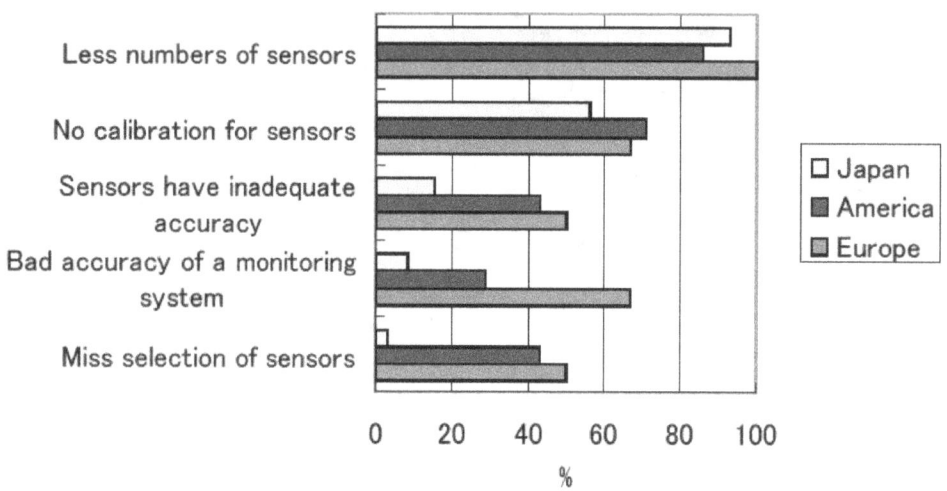

Figure 0-3. Dissatisfaction with sensors and monitoring systems in existing buildings

2. Do existing buildings have sufficient space for the installation of additional sensors?

Figure 0-4 shows survey results on machine room space. The percentage of buildings with enough machine room space to install additional sensors is 16 % in Japan, 52 % in America, and 39 % in Europe. Machine room space in America is large compared to that of Japan. Small machine room space poses a challenge for sensor installation in Japan.

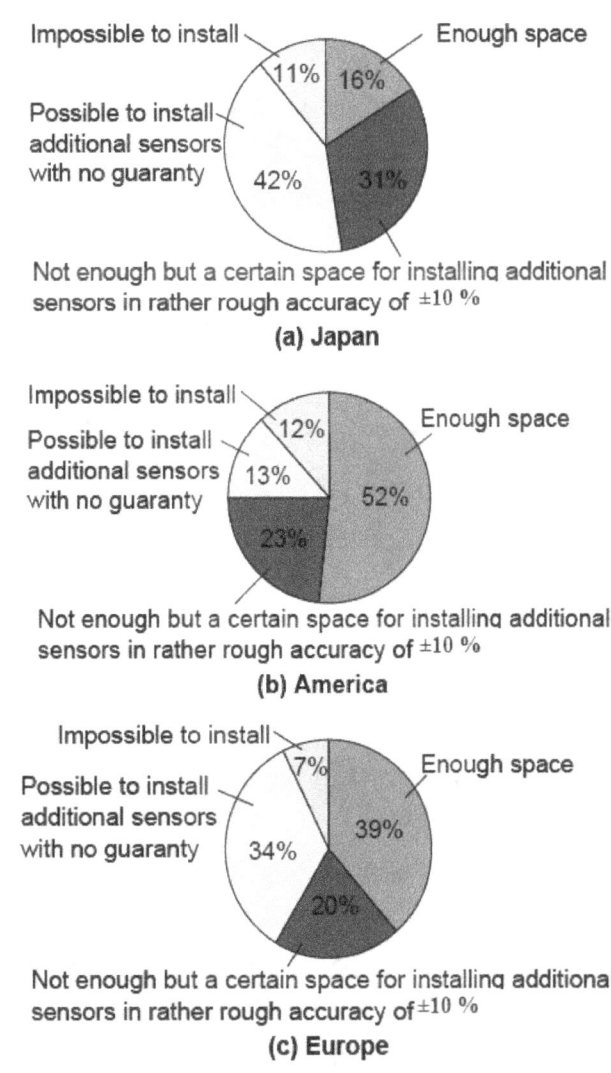

Impossible to install

Enough space

Possible to install additional sensors with no guaranty

11% 16%

42% 31%

Not enough but a certain space for installing additional sensors in rather rough accuracy of ±10 %

(a) Japan

Impossible to install

Enough space

Possible to install additional sensors with no guaranty

12%

13% 52%

23%

Not enough but a certain space for installing additional sensors in rather rough accuracy of ±10 %

(b) America

Impossible to install

Enough space

Possible to install additional sensors with no guaranty

7%

39%

34%

20%

Not enough but a certain space for installing additional sensors in rather rough accuracy of ±10 %

(c) Europe

Figure 0-4. Machine room space survey

3. Do affordable sensors in the marketplace have sufficient accuracy for purposes of energy management?

There were few valid responses concerning accuracy but there were several valid responses concerning sensor type.

Figure 0-5 shows the sensor types most commonly used in existing buildings. They are: resistance temperature detectors (water: 82 %, air: 66 %); thermo couples (water: 12 %, air: 25 %); ultrasonic water flow rate (56 %); electromagnetic flow (39 %); thermal air flow rate (39 %); Karman vortex (21 %); and Pitot tubes (11 %).

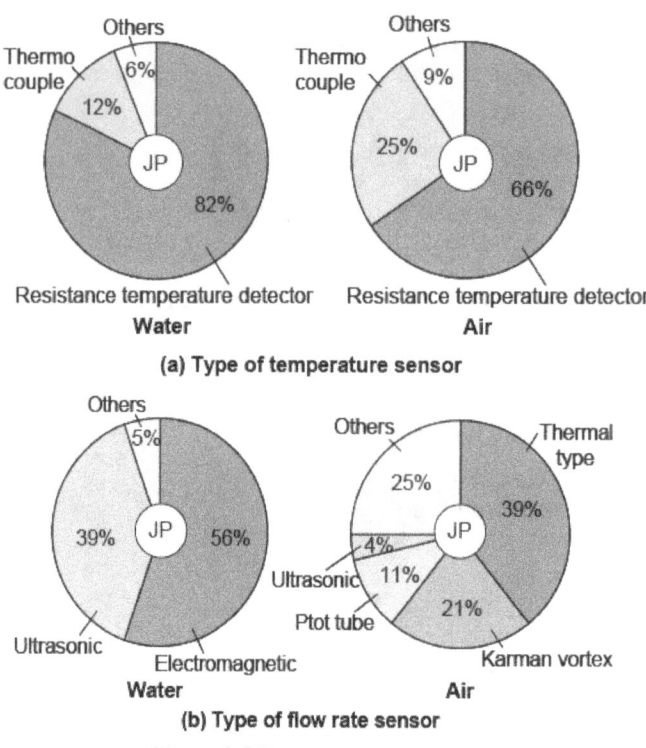

(a) Type of temperature sensor

(b) Type of flow rate sensor

Figure 0-5 Survey on sensor type

4. Do owners of existing buildings understand the need of additional sensors and allow for this in their budgets?

Figure 0-6 shows the survey results of owner reactions to budgeting. Most respondents in Japan answered that many owners completely agree with the Level A budget but there are some cases where owners agree with Levels C-F budgets. In most cases, owners seem to disagree with budgets of high level energy management systems in Japan.

Trends in America and Europe are similar to those in Japan but the ratio of American and European owners completely in agreement with Levels C-F are less.

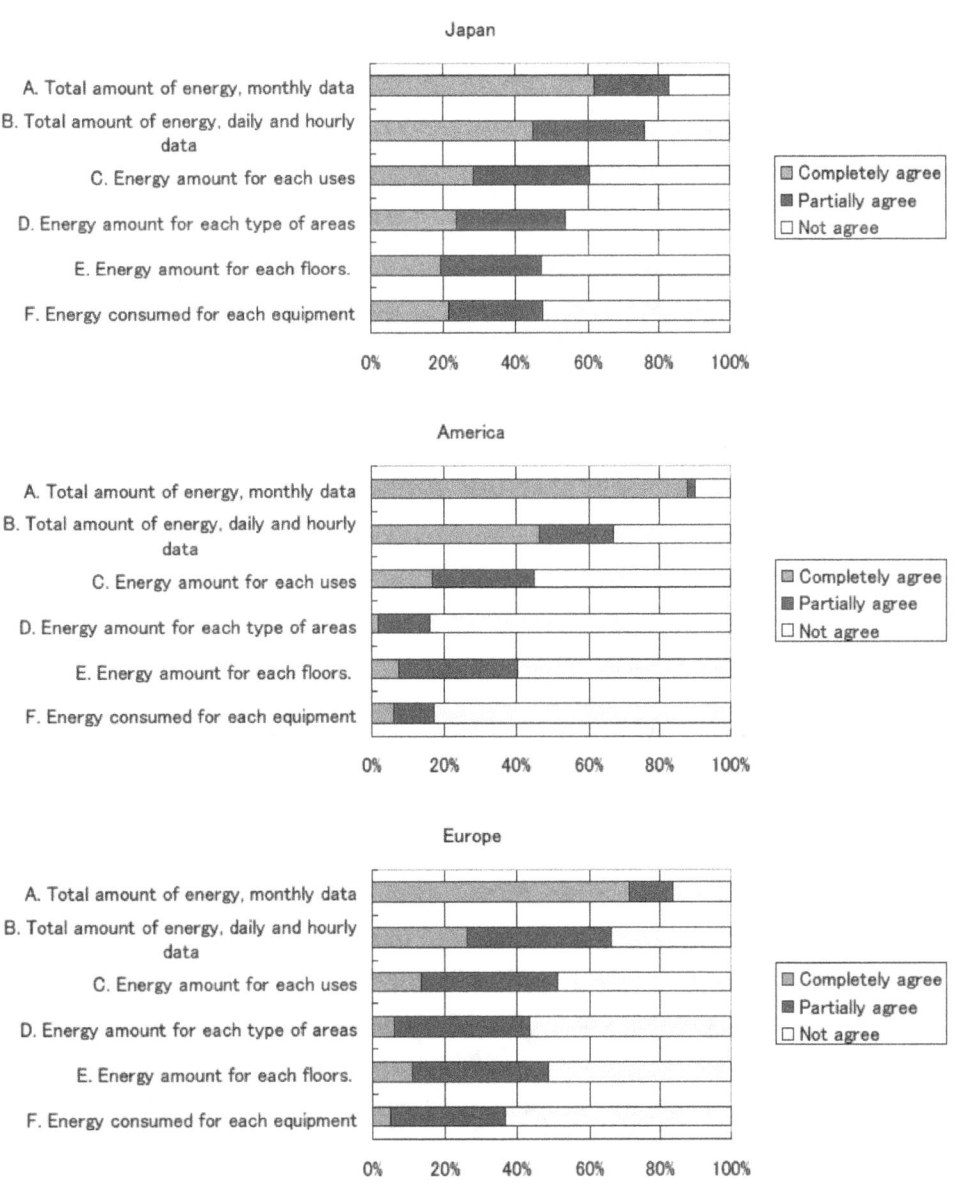

Figure 0-6 Owner reaction to budgeting

6.2   APPENDIX-2 Example of error in sensing system

Figure 0-7 shows a temperature sensing method that uses PT100 Ω (JIS C1604 class elements. The analog portion of the temperature sensing error is shown.

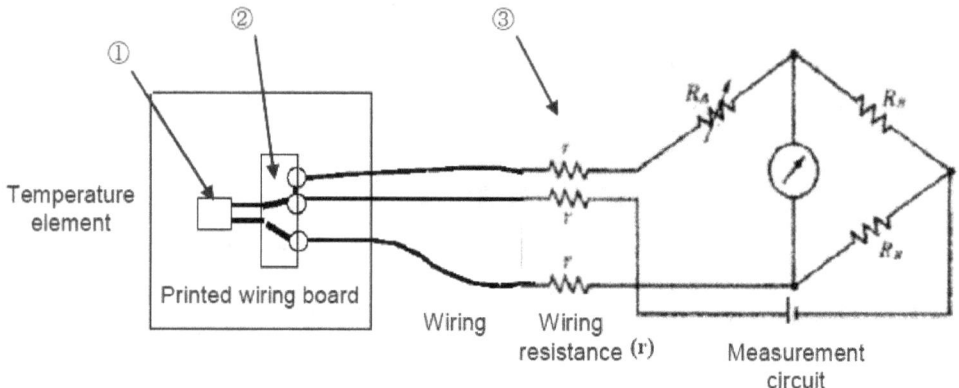

Figure 0-7 Indoor temperature sensing with temperature elements

1) Sensors

PT100 Ω (JIS C1604 class A)

The resistance of this element changes from 100.00 Ω to 123.24 Ω over a temperature range of 0 °C to 6 °C. The error is +/-0.15 °C at 0 °C. At 60 °C, there is a temperature dependence factor of 0.002*60=0.12, so the effective error at 60 °C is (0.15+0.12=) +/-0.27 °C. This is the value specified by JIS C1604 class A (2).

2) Printed wiring boards

The resistance in the wiring on the copper foil printed wiring board that carries the signal from the PT100 Ω element to the external pin is added as a positive resistance, which is effectively the measurement error. The value of the error calculated from the size of the copper foil, etc., is 0.04 °C. Thus, the overall error within the temperature measurement range (0 °C to 60 °C) is +0.04 °C.

3) Wiring

A bridge circuit that is incorporated on the instrument side, as the measurement circuit diagram shows, cancels out wiring resistance by using three-wire wiring. However, three-wire wiring is not entirely the same, so wiring error also exists here, and the error differs with the installation location.

In addition to the analog errors described above, sensing system error includes AD conversion error, which occurs in the conversion of the analog signal to a digital form. Furthermore, because the measurement results from two temperature sensors and the flow sensor measurement results are used to calculate the amount of heat, the errors in those measurements are included.

6.3   APPENDIX-3 Sensors Used by BEMS

The specifications of sensors used by BEMS (measurement range, accuracy, and output) are summarized in 'A Guide to Sensors for BEMS' in ANNEX 16 (p. 10). One sensor that is currently often used in performance evaluation but is not described in that document is the ultrasonic flow meter. Typical specifications for that type of meter are listed below.

Ultrasonic type

Accuracy: RD +-(1.0 to 3.0) %
Nominal diameter: (13 to 5000) mm
Flow range: (0 to 60, 000_ m3/h

6.4 APPENDIX-4 Features of EB and LEB Cx Tools Developed within Annex 47

TOOL NAME	DABO (Diagnostic Agent for Building Operation) Version2007
COUNTRY / ORGANIZATION	Canada Canmet Energy Technology Centre-Varennes
TARGETED BUILDING SYSTEM TYPE(S)	Institutional/Commercial
CONTACT PERSON:	Daniel Choinière

Objectives

The objective of the tool is to assist the building operator and commissioning providers in the early detection and diagnosis of faults and optimization of suboptimal operating parameters and sequence of operation in HVAC systems. These actions are often included as part of an ongoing commissioning process.

Table 0-2 summarizes the actions that the tool performs for fault detection and diagnosis at the component level, system level and energy and comfort performance auditing level.

Table 0-2. Functions of DABO^{TM}

Component Level – Fault Detection and Diagnosis
• Sensor: temperature, relative humidity, air flow, pressure, CO_2, current
• incorrect reading, complete failure
• Damper, valve, actuator
• stuck, incorrect position feedback, incorrect minimum position, defective leakage
• Coil, humidifier, fan
• insufficient capacity
• Controller
• logic, tuning, signal, unstable (set point, output)
• Set points: temperature, pressure, ventilation, relative humidity, CO_2
• Point in manual
System-Plant Level – Fault Detection and Diagnosis, Optimization
Helps building operators perform analysis for the following:
• Undersized and oversized components, systems and plants
• Non-optimum set points
• Temperature, pressure, humidity, ventilation, start-up time
• Inappropriate sequences of operation and schedules
• Poor energy management and peak load control

Energy, Operation Quality Control and Comfort Performance Report
• Energy and comfort profile (hour, day and month)
• Room level, equipment level (terminal unit device, AHU, heating/cooling power plant)
• Building level (electric, gas and oil meter)
• Historical database for energy audit or retrofit
• Monitoring of energy saving measures

Functions

Functions of the tool are described using IDEFØ. Figure A01-1 shows the IDEFØ diagram of the tool. The tool monitors measurement and control point data and analyzes it to detect symptoms of abnormal behavior in various HVAC components, such as un-calibrated or failed sensors, actuator or linkage failures, controller instabilities, non-optimal sequences of operation, etc. The tool can also diagnose the possible causes and provide explanations for abnormal behavior. It goes beyond the capabilities of conventional BEMS alarms by utilizing the same knowledge and reasoning that an expert building engineer would apply to detect problems. The tool analyzes all systems continuously (24 hours a day, 365 days per year).

The tool also produces reports adapted to the different professionals involved in the operation of buildings (building operators, service technicians, energy managers, commissioning agents, HVAC&R engineers).

Applied in an ongoing commissioning process, the functions of the tool are:

During the investigation phase :

- Help to detect where and when there is a problem

    - Identification of potential problem
    - Identification if installation meets occupant needs and if installation type corresponds to building utilization

- Is optimal sequence of operation implemented?

    - Identification of basic optimal sequence and low payback energy saving measures to check

During the implementation phase :
- Use to check if problem persists
- Used to check if new control strategy performs as intended
- Use to see new energy efficiency measures

During the hands-off and persistence phase:
- Use DABO periodically to detect faults that may prevent operation of optimal control strategy
- Help perform tasks on the Persistence Check List to ensure performance

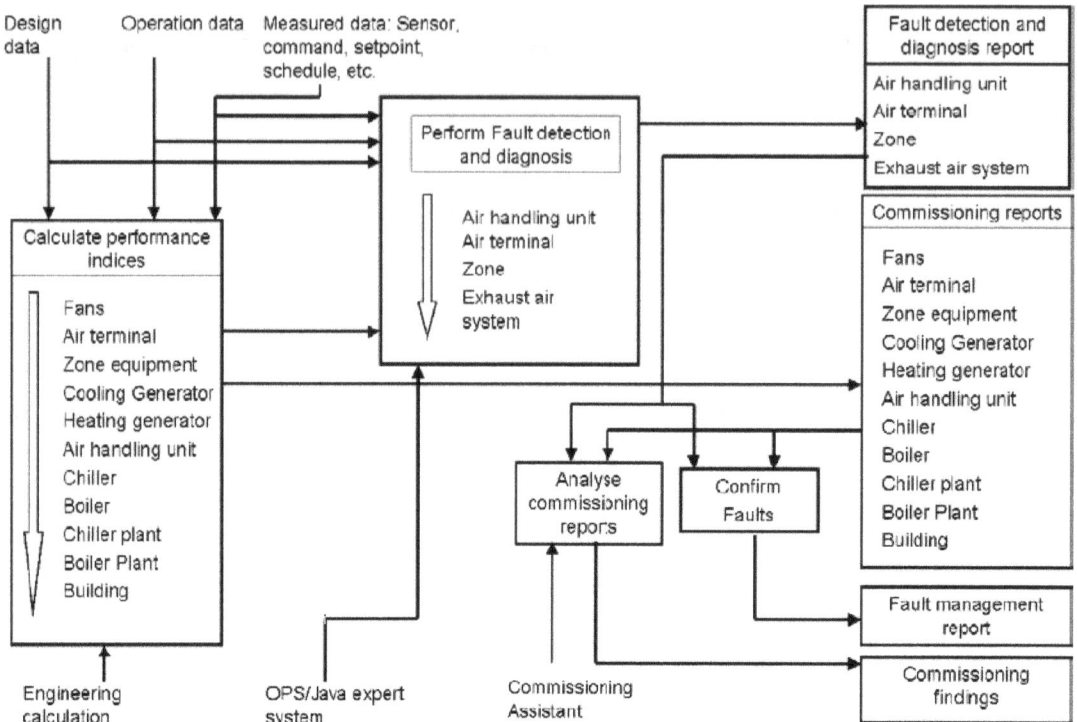

Figure 0-8. IDEF0 diagram of DABO

Data Management

See Appendix 5 for details on required Input Data, Design data, and Operation data.

Measured data: Control point values (Sensor, command, setpoint, schedule, etc.) from the BEMS

Output: DABO produces four types of reports:

FDD report displays data generated by the various fault detection services. Data are shown in a colored table.

Commissioning report (COMM) displays data generated by the various commissioning services or data from the building. Data are shown in a colored table. This report can be configured by the user for additional statistical analysis.

Points report displays data from building points selected by the user. Data can be shown as a line chart or a table. This report can be requested from the FDD report, from the commissioning report, or from the "Tools" menu.

Fault management report (FMR) displays confirmed and repaired faults by device type.

Implementation

DABO™ resides on a Personal Computer and analyzes incoming data from the BEMS. It is compatible with most existing building control systems.

DABO™ was designed using UML standards in 'Together' environment. The DABO platform was developed in JBuilder and Together environments (Borland Enterprise Studio 7 for Java) in the JAVA programming language.

Excelsior v4.1 software was also used to create Windows (2000 and XP professional) executables from Java machine dependent packages.

The fault detection expert rules are coded with an expert system programming language called "OPS/JAVA".

The connectivity to BEMS application is coded in C++.

DABO™ uses a relational database usually managed on an SQL 2000 database server or equivalent.

Operability

DABO's real-time optimization service analyzes data at three levels:

At the first level, an automatic component analysis of individual HVAC devices and equipment is performed using current commissioning and FDD procedures to verify their proper operation. These tests are performed each hour. This level uses a combination of calculated performance indices and a Rule Based Expert System to perform the analysis; further explanations are provided in the following sections. The DABO reports for the Fault Detection and Diagnosis Component analysis are called FDD reports.

The second level of testing consists of an integrated semi-automatic system analysis to verify the operation and energy performance of the overall HVAC system over a longer period of time (e.g., hours, days, weeks or months). This level uses a combination of calculated performance indices, graphical statistical analysis and HVAC system optimization reference tables to help users of the tool to perform an ongoing commissioning analysis of the installation. Further explanations of the performance indices, statistical analysis and reference optimization table are provided in the following sections. The DABO reports for the fault detection and diagnosis (system and plants) analysis are called commissioning (COMM) reports.

On the third level basic energy, comfort and operation control quality auditing is performed. This step uses a combination of calculated performance indices, graphical statistical analysis and user defined report structures to audit the installation. It also helps to follow energy saving measures and to provide full building historical data that could be used in various detailed energy audits and in the development of HVAC and lighting system optimization concepts. The DABO reports for the energy, operation control quality and comfort performance analysis are also called commissioning (COMM) reports.

Analytical Engine

The tool is based on a combination of artificial intelligence algorithms (Rule Based Expert System), calculation of performances indices and graphical statistical analysis.

End Users

The tool is designed to be used by building operators, service technicians, energy managers, commissioning agents, and HVAC&R engineers as a function of their respective needs. (fault detection, system optimization, performance evaluation, ensure persistence of commissioning and energy savings measures benefits)

Benefits

Benefits of DABO within the ongoing commissioning process are:

General:

- 15 % to 30 % energy savings
- Improved comfort and reduction of complaints
- Facilitate maintenance
- Ensure Persistence in energy savings measures
- Extend HVAC equipment life

During Investigation and Implementation:

- Encourage widespread application by reducing its cost
- Improve the quality assurance process
- Allow application during entire life of the building

During Hands-on and Persistence of Benefits

- Help to structure the continuous process
- Optimize the periodic check list verification
- Improve the quality assurance process
- Reducing persistence cost
- Keep historical Data

Target Systems to be tested

The tool is actually tested in 10 recommissioning projects (demonstration) across Canada. It is applied for air- handling unit, air distribution system, heating and cooling network and energy management strategies. Demonstration projects include office buildings, a warehouse, a convention center and a courthouse.

No. 02   SUBTASK B: Commissioning and Optimization of Existing Buildings

TOOL NAME	CITE-AHU
COUNTRY / ORGANIZATION	USA/ National Institute of Standards and Technology
TARGETED BUILDING / SYSTEM TYPE(S)	Air Handling Units
CONTACT PERSON:	Natascha Milesi Ferretti

Objectives

CITE-AHU is a software program designed to facilitate the analysis and maintenance of air handling units in office buildings. Operational faults for AHUs in office buildings can negatively impact energy consumption, maintenance costs, thermal comfort for the occupants and environmental protection.

CITE-AHU aims to facilitate the monitoring of AHUs and provide the ability to detect faults related to improper operation, using the APAR algorithm (expert rules), to present results in a hierarchical way with a list of likely fault causes, and to assist in documenting results, providing electronic reporting templates.. The goal is to optimize the system performance in terms of occupant comfort and energy management. It was conceived exclusively for tertiary buildings (i.e., offices, shops, hospitals) equipped with one (or several) AHUs. The fault detection method developed can be applied to several types of single-duct air-handling units.

Functions

The tools main function is to detect faults in the air-handling unit. Non-standard features include heat recovery, humidifier, and economizer applications. Active testing, injecting test signals to evaluate system responses under various conditions, are carried out using BACnet. Passive monitoring, using BEMS data, can be performed regularly (daily, weekly, or monthly).

The types of faults that can be detected are:

Mechanical failures such as failed/drifting sensors, stuck/failed actuators, duct blockage, leaking valves/dampers, slipping/broken belts, etc.,

Control problems such as PID loop tuning, inappropriate setpoints, manual overrides, and incorrect sequences/programming,

Degraded performance including heat exchanger fouling,

171

Design faults such as undersized duct/piping, mismatched components, inappropriate zoning.

Data Management

Data is obtained directly through the BEMS. Typically 1 min to 10 min data readings are needed. This data can be obtained using the BACnet Data Source, a NIST-developed program that facilitates data collection with and BACnet enabled devices. The time-series data is collected in CSV files with columns for the date and time stamp, along with the requited input data, presented below.

Input Data

- Supply, return, mixed, and outdoor air temperature
- Outdoor air humidity and/or return air humidity and/or enthalpy (for enthalpy-based economiser control
- Heating coil valve signal, cooling coil valve signal, damper control signal

Output Data

A graphical user interface presents weekly summary of data analysis results, providing users with the ability to drill down for greater detail and additional visualization tools

In addition to the graphical display, fault reports are generated and can be archived. In Active testing, control signals and setpoint changes are made directly to the controllers using the BACnet Data Source.

Training Data

No training data required, but expert knowledge of the system is needed to configure the tool (control sequences, fault thresholds) for the application.

Implementation

CITE-AHU resides on a personal computer and analyzes the data from the BEMS using imported CSV files. In its current version, developed in Visual Basic, expert rules are incorporated into the software.

The BACnet Data Source used to facilitate data collection is implemented in C and C++. It is implemented by user who creates test sequences or scenarios into the Active Test Module of Cite-AHU. The Active Test Module is designed to inject test signals directly to the controllers. Communication is established with the BACnet dll by sending request parameters in an array, and the command is then sent to the AHU.

CITE-AHU software uses the control signals to detect the specific operating mode. For each operating mode of the air handling unit, expert rules have been developed in order to detect common operational faults. Each time that the conditions of an expert rule are satisfied, faulty operation is identified. In the event of awkward repetition of certain faults, the software allows the user to deactivate the faults for each operating mode. Each fault corresponds to one or several fault causes. The software posts the cause the most often detected and thus the most probable.

There are 8 operating modes:

Heating: the supply air is heated by the heating unit (this air is composed of X % of fresh air and (100-X) % of return air);

Free cooling: fresh air is mixed with return air and is neither heated nor cooled mechanically;

Mechanical cooling with 100 % of outdoor air: the supply air composed of 100 % of outdoor air and is cooled by the cooling coil unit;

Mechanical cooling with a minimum of outdoor air: the supply air composed of a minimum amount of outdoor air and is cooled by the cooling unit;

Night cooling: this is for free cooling during unoccupied periods when temperatures permit supply air;

Frost-protection: this operating mode of the air handling unit is switched on when outdoor air is too cold in order to protect the installation;

Stop: the air handling unit is stopped; and

Unknown: this operating mode is detected when none of other operating modes are activated and detected.

Each operating mode has its own rules of operation. When the operating data is evaluated, faults are detected by the software. Furthermore, because in many cases, there are multiple causes that can cause a specific fault, a list of possible fault causes are presented to the user.

If certain faults keep recurring, the software allows the user, for each operating mode, to deactivate the detection of faults which you no longer wish to observe (i.e., if a service call has been scheduled to address a specific problem

Steps followed by the CITE-AHU software:

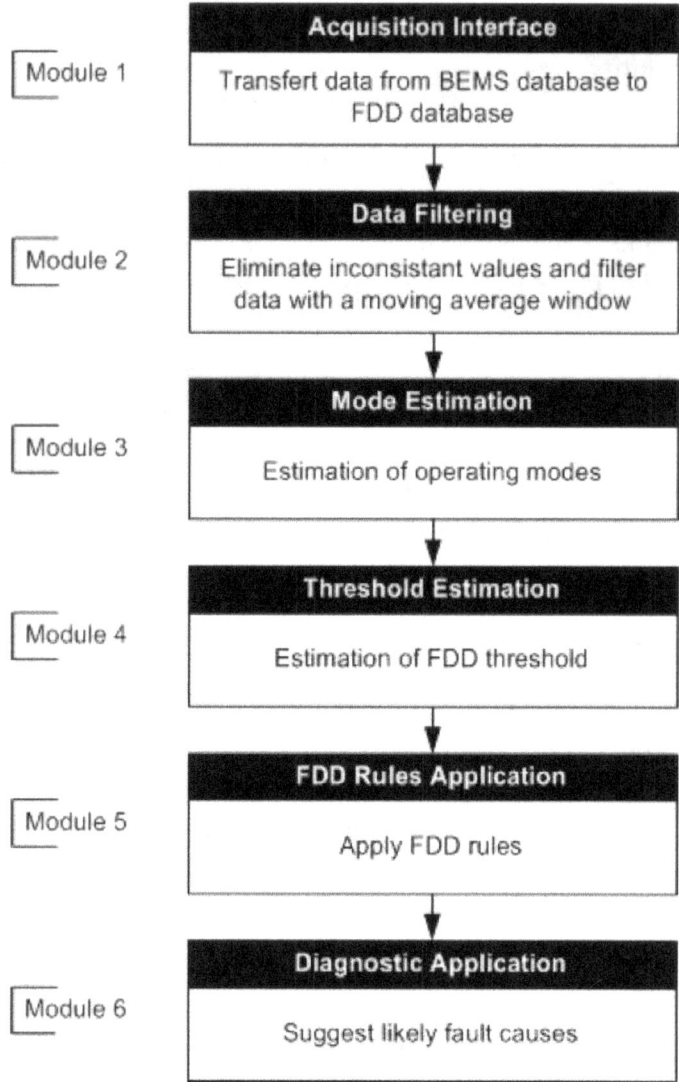

Figure 0-9 Steps followed by CITE-AHU

Analytical Engine

The FDD tool (Visual Basic) is a rule-based expert system, using principles developed by NIST.

End Users

The primary users of the tool will be (1) commissioning agents performing tasks to verify correct installation and operation, and (2) building operators/ energy managers who want to monitor performance of multiple AHUs on an ongoing basis, performing seasonal pre-functional checks

CITE-AHU aids building operators in decision-making by analyzing data collected from the building energy management system (BEMS) and evaluated using a set of expert rules governing the proper operation of the mechanical equipment. Thus, CITE-AHU software has many advantages in particular:

- It can analyze large batch files (which is processed by means of computer);
- It analyzes and preprocesses information collected from the BEMS;
- It provides fault detection of the installation by applying expert rule analysis;
- It presents results simplified in the forms of graphs and tables; and
- It provides a diagnosis by writing a list of possible causes of each fault detected.
- It also determines investigation priorities thanks to a system of hierarchical organization of the faults.

Target Systems to be Tested

DESCRIPTION OF TWO AHU CONFIGURATIONS

The first AHU layout, shown in Figure 0-10, is a constant air volume system with the following components:

- ☐A constant flow air supply fan supplying the terminal units via a single duct air distributionnetwork to provide conditioned air to the various building zones;
- ☐An exhaust fan to extract air from the occupied zones
- ☐Fresh air, mixed and exhaust dampers to regulate the flow of air supplied to the building;
- ☐A heating coil unit to heat the supply air
- ☐A cooling coil unit to cool the supply air;
- ☐A filtration unit to trap particles in the air and
- ☐A humidifier to regulate the relative humidity of the supply air.

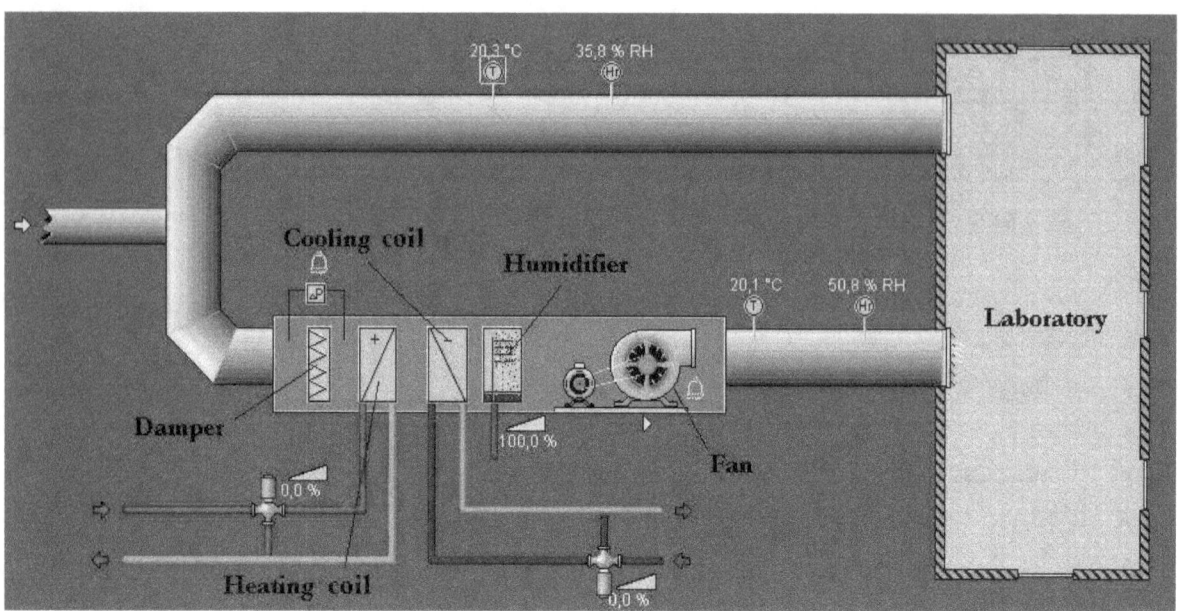

Figure 0-10 Layout for AHU 1

The second AHU layout, shown in the Figure 0-11, is a variable air volume system with the following
components:

- □A variable flow air supply fan supplying the terminal units (VAV unit: Variable Air Volume)via a single duct air supply network to distributes the air to the various parts of the building;
- □An extractor fan to extracts air from these areas;
- □Fresh air, return and mixed and exhaust registers to enable part of the air extracted from the building to be used in order to heat up the delivery air;
- □A heating coil unit to heat the delivery air, and
- □A filtration unit to traps particles in the air.

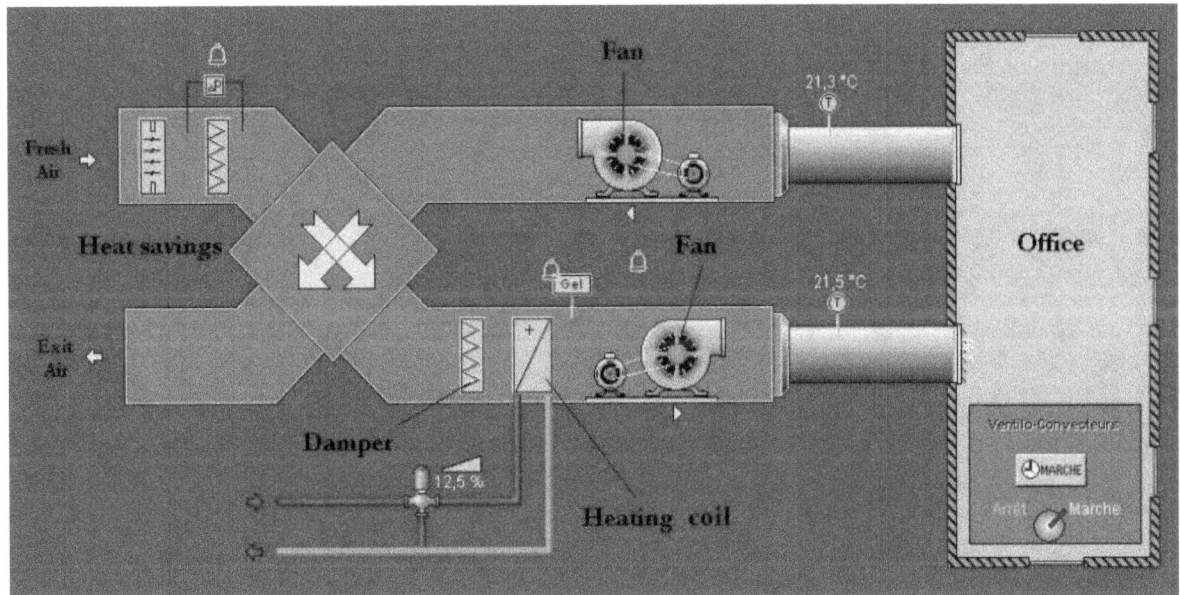

Figure 1-11 Layout for AHU 2

Case Studies (Completed)

Korean Institute for Energy Research, Taejon, Korea, 3 AHUs

Centre Scientifique et Testing Du Batiment (CSTB), Paris, France, 2 AHUs

NIST whole building emulation w/ real controllers, Gaithersburg, MD, USA, 2 AHUs

Case Studies (Planned)

NIST Campus, Gaithersburg, MD, USA, 6 AHUs

No. 3 SUBTASK B: Commissioning and Optimization of Existing Buildings

TOOL NAME	Proto-type FDD tool "i-BIG" for HVAC systems
COUNTRY / ORGANIZATION	The Netherlands / TNO Built Environment and Geosciences
TARGETED BUILDING / SYSTEM TYPE(S)	Any type of buildings / Air handling Units
CONTACT PERSON:	Henk Peitsman, Lucienne Krosse

Objectives

The tool is for fault detection and diagnosis (FDD) of HVAC systems, regarding the operation of the HVAC system and also to review the operation conditions of HVAC components to enable component and system level optimization, which will be termed continuous optimization. The tool is installed in a PC, (i) connected to the data monitoring network of the BEMS or (ii) by remote control. The use of the current tool supports keeping air-handling units (AHU's) maintaining normal status constantly. The tool helps detect faulty situations.

Functions

The function of the tool is to detect faulty places through the whole AHU system. Most component level faults and some system level faults can be detected using the tool. The tool can maintain optimal performance of the AHU by ongoing commissioning (ongoing monitoring).

Examples of faults: (i) faults in system control, (ii) faults in schedules of operation, and (iii) design errors (e.g., undersized or oversized equipment), and (iv) inappropriate room temperature settings by occupants.

Data Management

Data from existing BEMS must be used whenever possible. Data of 10-minute intervals or less are necessary to detect fault or poor performance operation. If no data from BEMS are available, assumptions must be made concerning (i) operation of HVAC, (ii) threshold values. The measured time-series data is stored in CSV files, which have a defined time label in each row. The time label indicates year, month, day, day of the week, hour, minute and second.

Input Data

- Weather data;
- Room temperature and humidity;
- Inlet/outlet water of heating and cooling coils;
- Supply air temperature;
- Schedule of operation;
- All type of set points; and
- Mass flow rates, if available.

Implementation

i-BIG resides on a personal computer and analyzes the data from the BEMS. In its current version, expert rules are incorporated in the software. The i-BIG tool is developed in Visual Basic.

Operability

A user interface is developed based on continuously monitoring input data of the HVAC system. It shows graphical output of sensor and control signal as a function of time, and associated fault conditions and diagnoses. System faults or troubles, including possible causes or inappropriate operation, are presented. Trend curves are presented on the screen. Operation vizualisation module presents (i) analog input sensor signals and (ii) control signals.

Analytical Engine

The FDD tool (Visual Basic) is a rule-based expert system, using principles developed by NIST.

End Users

The primary end users of the tool will be (i) service companies responsible for HVAC system maintenance. They will use the component level diagnostic information to assist them with troubleshooting and to prioritize maintenance activities; (ii) building operators/energy managers who want to gauge the performance status of the HVAC systems; (iii) Re- commissioning providers can also use the tool to prevent faulty operation and ensure good performance of the HVAC systems.

Benefits

Benefits of i-BIG use in ongoing commissioning process:

- Improved comfort and reduction in complaints;
- Energy savings: (10 to 35) %;
- Facilitate maintenance and service companies;
- Extend HVAC equipment lifecycle;
- Optimize periodic check list verification; and
- Keep historical data.

Target Systems to be Tested

The tool is actual tested in two projects to demonstrate its benefits for AHU system FDD. Data is automatically collected in our laboratory from an AHU in a swimming pool (remote). A second project is executed at a hospital AHU.

TOOL NAME	Performance Analysis Tool for Heating System
COUNTRY / ORGANIZATION	Norway/ Norwegian University of Science and Technology
TARGETED BUILDING / SYSTEM TYPE(S)	Heating system
CONTACT PERSON:	Natasa Djuric

Objectives

The tool is aimed at heating substations in the district heating supply. It is for use at the system level to estimate performance. The operators usually assumes causes of faults in the case of a problem. With this tool the answer can be found systematically. The tool is configured to estimate abnormal indoor temperature and gauge if there are some biases in the set parameters. In addition, it can estimate changing energy consumption due to bias. The tool can assist in determining faults in the heating system.

Functions

The tool consists of two parts: the calculation part and the fault diagnosis part, shown as 2D plots. Tool functions are described using IDEFØ in Figure 0-12. The calculation part consists of two modules: the heat balance model, which is shown as Activity C1 and C3 in Figure 0-121, and the optimization module, which is shown as Activities C2 in Figure 0-12. The fault diagnosis part is shown as Activity C4.

Data Management

Figure 0-12 outlines the information required for application of the tool using the IDEFØ diagram. The input data on the left are used as hourly data. Index 1 beside input data in Figure 1 implies that this data set is used for model calibration, while Index 2 implies data used for fault detection. The input data are obtained by additional measurements (like outdoor temperature and tap water energy consumption) and logging in BEMS. The output data are used for 2D plots.

Implementation

The calculation part of the tool is developed in MATLAB and consists of two modules: heat balance model and the model calibration/optimization module. The tool is configured using BEMS data and on-site measurements. In addition, approximate values of the building envelope characteristics and radiator properties are required.

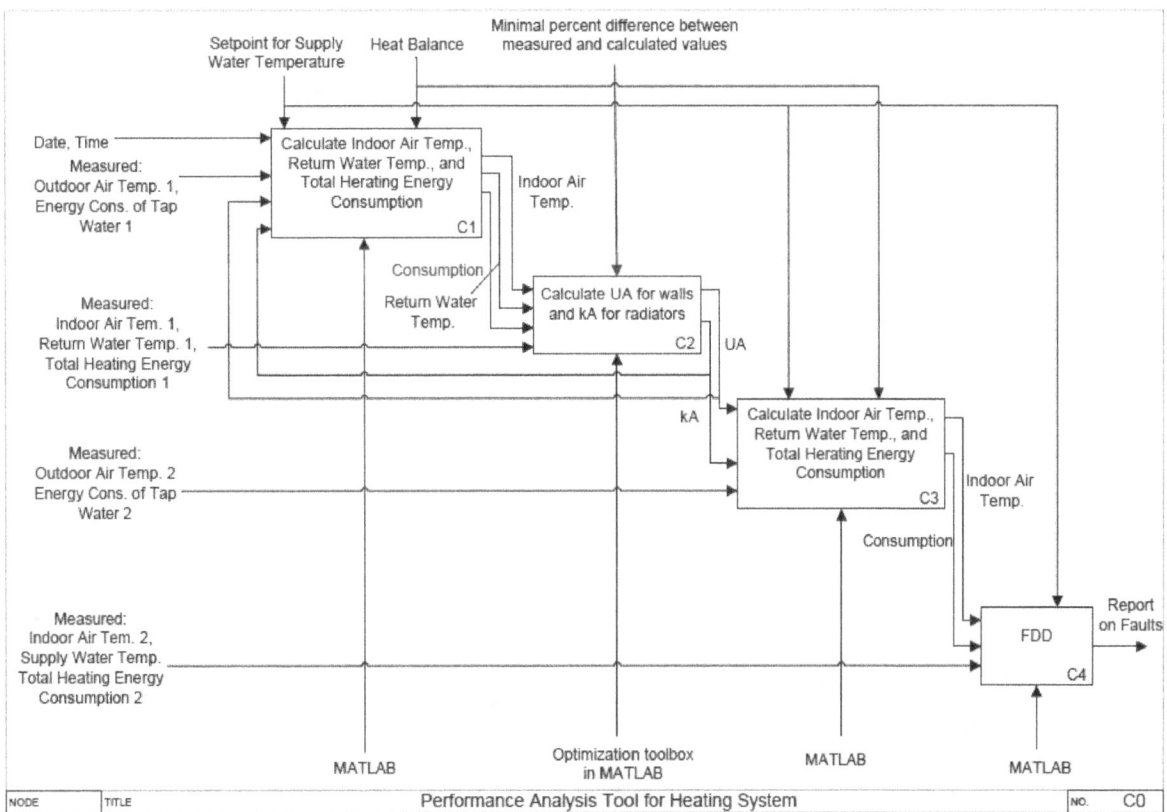

Figure 0-12 IDEF0 tool diagram

Operability

The tool output is of practical use to building operators. Figure 1 shows the calculation part that is research work, while rules for fault diagnosis, the tool output, are of practical use to building operators. Tool outputs are given as 2D plots both for correlation and time-series. This way, fault diagnosis rules are generated by detecting biases in performances.

Analytical Engine

The tool is based on the simple heat balance model. The model is calibrated by comparing the measured data to model output data for each time step. The system model is developed in MATLAB. The calibration is performed using a sequential quadratic programming algorithm.

End Users

The tool is designed to be used by building operators.

Benefits

It is difficult to get a model that performs similarly to a real system. By using an optimization algorithm a simple model can be easily calibrated against real data. When the model is established, it is simple to utilize it for fault detection by comparing the model outputs to actual performances.

Target Systems to be Tested

The tool was tested on the Material Technology Building on the campus of the Norwegian University of Science and Technology in Trondheim. Data collection and logging was easily supported by the building operators. Afterwards, results and fault causes were easily understood when presented in 2D plot formats.

No. 5   SUBTASK B: Commissioning and Optimization of Existing Buildings

TOOL NAME	[not yet defined]
COUNTRY / ORGANIZATION	Germany / Fraunhofer Institute for Solar Energy Systems
TARGETED BUILDING / SYSTEM TYPE(S)	Whole Building (non-residential)
CONTACT PERSON:	Christian Neumann

Objectives

The tool can be used and supports a top-down analysis approach for the evaluation of building performance. To date, tool features are: predefined visualization based on a minimal data set and automated outlier detection for energy and water consumption (see next paragraph). The tool is used for manual fault detection and diagnosis based on visualization, followed by continuous monitoring using the outlier detection feature.

Originally, the idea was to couple the tool wherever possible with the EPBD but as a result of the actual state of the EPBD, linkage between commissioning and EPBD is difficult.

Functions

The analysis features offered in the tool are:

- Predefined visualization (time series, scatterplots, carpetplots) based on a minimal data set (see below).
- Visualization is based on a minimum data set with a unified point naming convention. That is, as soon as the data point names are defined the visualization is performed automatically, including data filtering and grouping.

- Outlier detection on energy and water consumption. This feature is also based on the minimal data set. It identifies the energy signature based on daily energy consumption values, taking into account different day types. After training it can be used for outlier detection.
- A feature still under development is model-based analysis (simplified dynamic) that calculates reference values for daily energy consumption.

The minimal data set is recorded every (5 to 10) min on an hourly basis

Data Management

The data (measured data and meta data) is persistently stored in an HDF 5 file. The data is structured within the file according to project, subproject, subsystem, sensorgroup, sensor. Charts and analysis results are also stored in the data. This allows different user interfaces (like webfrontend, GUI, console) without implementing functionality every time.

The minimal data set recorded comprises: total delivered energy consumption, total water consumption, weather data, temperature and humidity of reference rooms, system temperatures of main loops (air and water), major AHU exhaust and supply air moisture.

Implementation

The software is developed with python (www.python.org), an open-source scripting language. To implement functionality, several existing libraries (open source) will be used (like matplotlib for generating charts, scipy for numerical operations, etc.)

Statistic environment R (www.r-project.org) will be used for statistics analysis.

Operability

For our projects, results are presented as a web-based service (see: www.modben.org and www.buildingeq.eu). Data and visualizations are accessible there. A GUI is under development(May 09).

Analytical Engine

A robust linear regression approach combined with clustering and standard optimization routines are used for statistics analysis (identification of energy signatures, detection of outliers). The model based approach developed now is based on CEN standards

End Users

As the manual FDD based on visualization requires system knowledge, the end users to date are trained operation staff. We hope to automate at least part of the analysis.

Visualization:

Easy to apply, applicable to nearly every building, automated data processing and information generation

Automated oulier detection:

Quick detection of abnormal energy/water consumption.

Target Systems to be Tested

Whole building/Overall building performance. 9 demonstration buildings in Germany and eight additional buildings throughout Europe.

No. 6   SUBTASK B: Commissioning and Optimization of Existing Buildings

TOOL NAME	Optimization Tool for Air-Conditioning System Operation Considering Thermal Load Prediction Errors
COUNTRY / ORGANIZATION	Japan / Kyushu University
TARGETED BUILDING / SYSTEM TYPE(S)	
CONTACT PERSON:	Yasunori Akashi, Daisuke Sumiyoshi

Objectives

This tool is intended for on- going commissioning and be applied mainly to commercial use central air-conditioning systems.

Functions

This tool optimizes energy consumption and the indoor environment by changing set values of the HVAC&R system.

Data Management

Input Data

- Weather data (outdoor temperature and humidity, general weather conditions)
- Room temperature and humidity
- Inlet/outlet water temperature of chiller and coil
- Supply air temperature

- Schedule of operation

- These data can be obtained from BEMS.

Output Data

Optimized set-points concerning the following:

- Room temperature and/or humidity;

- Inlet or outlet water temperature of chiller and coil; and

- Supply air temperature, etc.

| Implementation |

The analysis is carried out in the following manner.

1. One day after weather data are predicted from current weather data.

2. Prediction load data are made.

3. The air-conditioning system simulation is carried out using prediction load data.

4. Optimize set-points by repeating step 3. A genetic algorithm is used for the optimization method.

| Operability |

Not Defined

| Analytical Engine |

FORTRAN

| End Users |

This tool is included in BEMS and can be operated automatically. The Building operator manages its operation.

| Benefit |

This tool can optimize the operational method considering thermal load prediction errors.

| Target Systems to be Tested |

The HVAC&R experimental analysis building in Kyushu University campus will be used to verify the tool prototype. We are planning to use one more office building for a case study(TBD).

IDEF representations of the tool are shown in Figures 0-13, 0-14, 0-15, and 0-16.

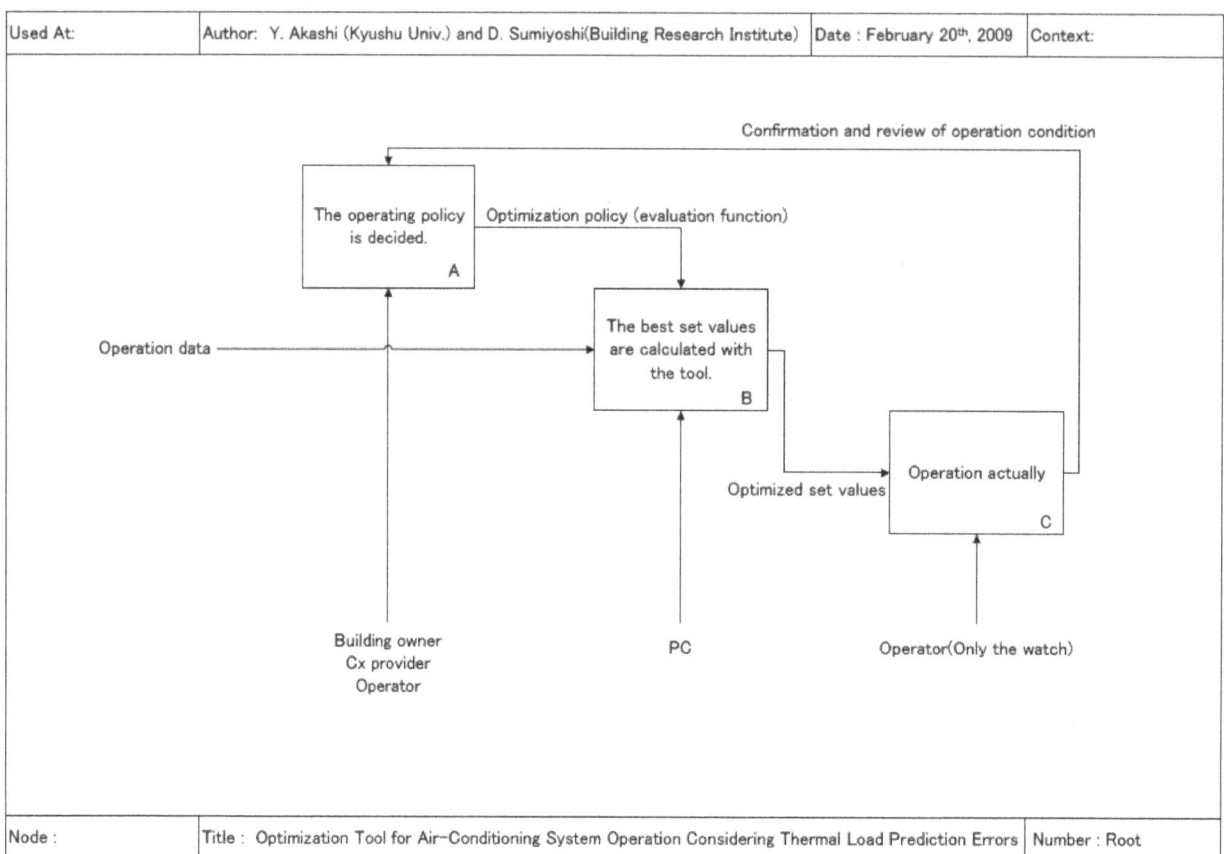

Figure 0-13 IDEFO tool diagram (Root)

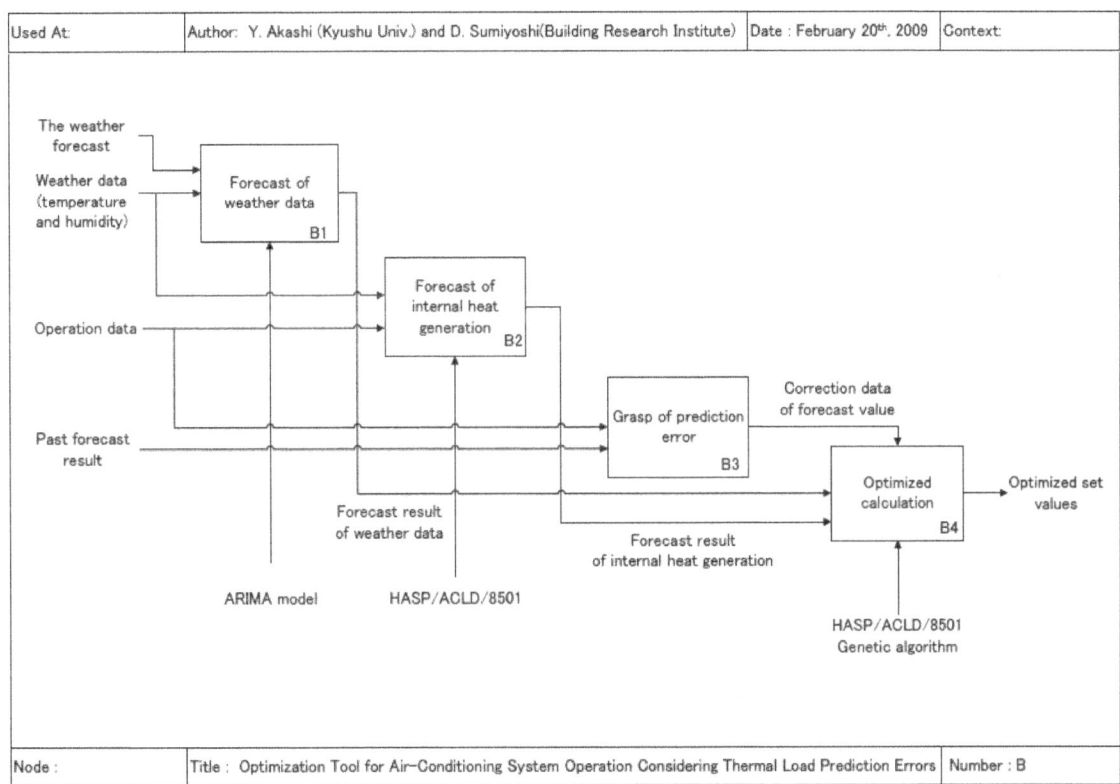

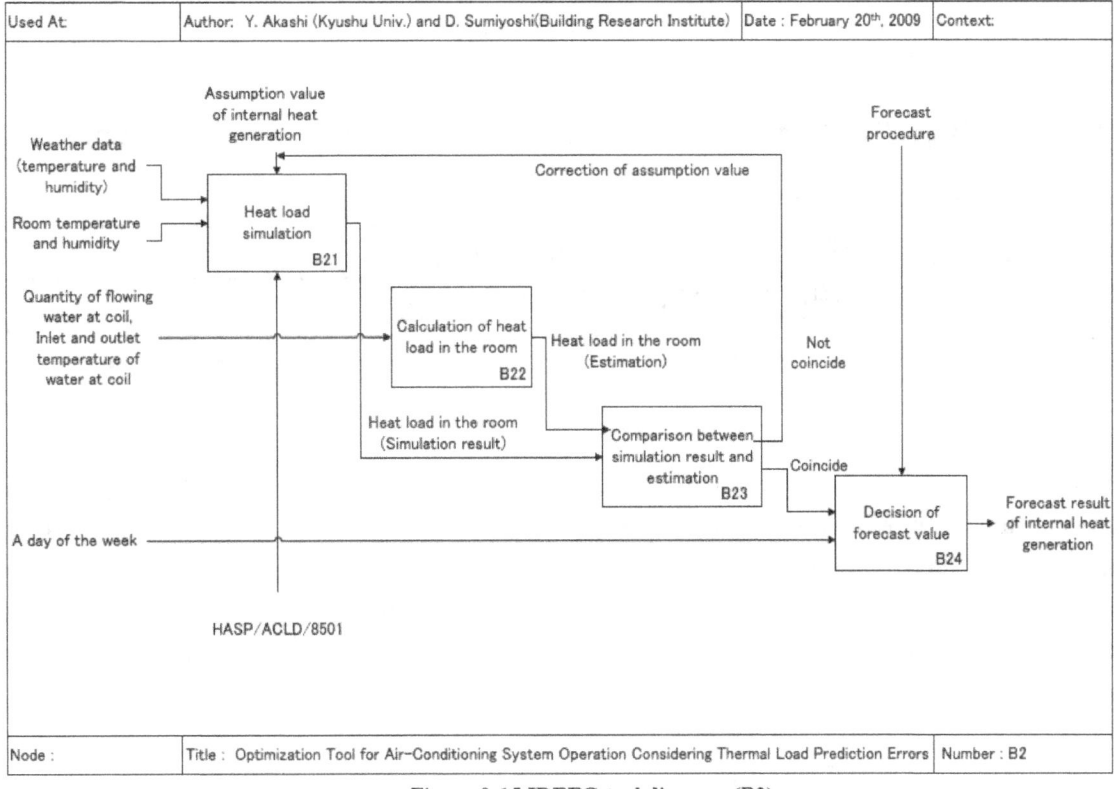

Figure 0-15 IDEF0 tool diagram (B2)

187

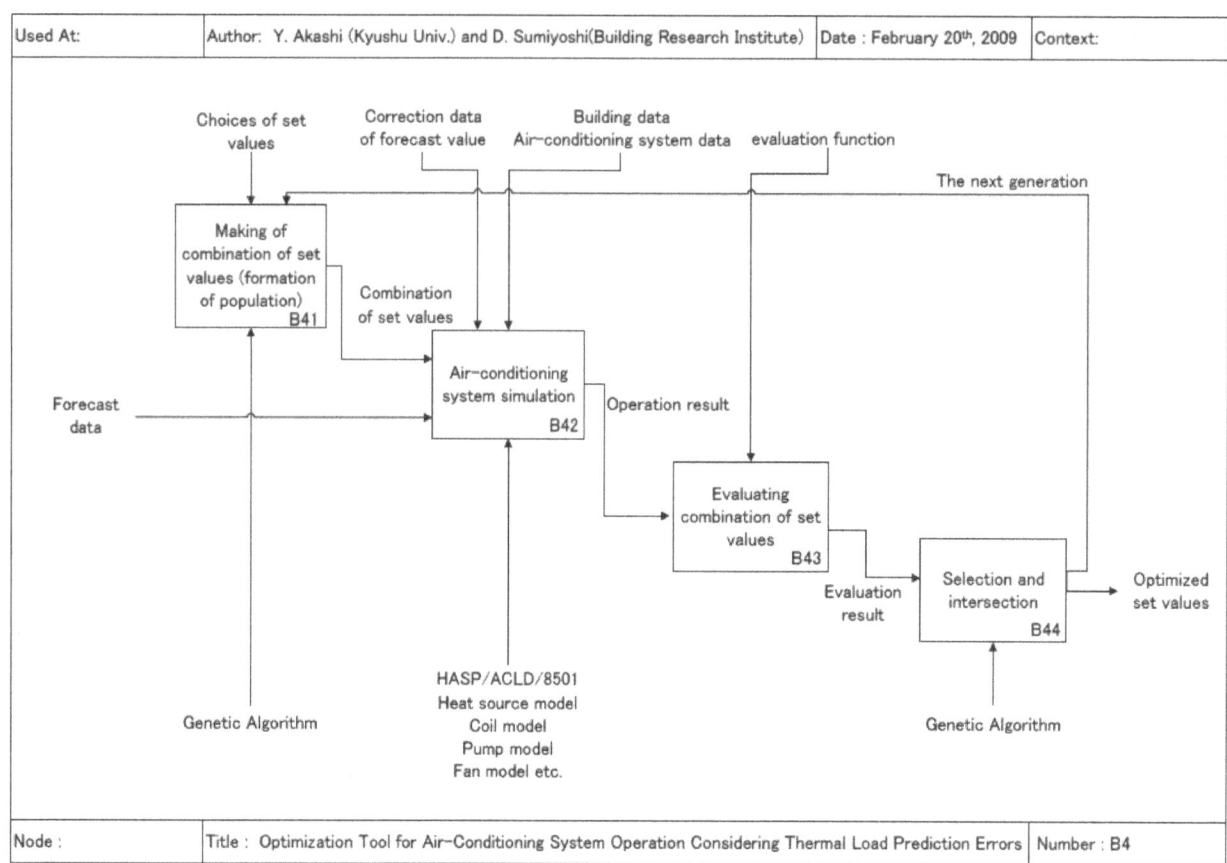

| Used At: | Author: Y. Akashi (Kyushu Univ.) and D. Sumiyoshi(Building Research Institute) | Date : February 20th, 2009 | Context: |

Choices of set values

Correction data of forecast value

Building data
Air-conditioning system data

evaluation function

The next generation

Making of combination of set values (formation of population) B41

Combination of set values

Forecast data

Air-conditioning system simulation B42

Operation result

Evaluating combination of set values B43

Evaluation result

Selection and intersection B44

Optimized set values

Genetic Algorithm

HASP/ACLD/8501
Heat source model
Coil model
Pump model
Fan model etc.

Genetic Algorithm

| Node : | Title : Optimization Tool for Air-Conditioning System Operation Considering Thermal Load Prediction Errors | Number : B4 |

Figure 0-16 IDEFO diagram of the tool (B4)

No. 7   SUBTASK B: Commissioning and Optimization of Existing Buildings

TOOL NAME	An ongoing commissioning tool for VRV package systems
COUNTRY / ORGANIZATION	Japan / Chubu University
TARGETED BUILDING / SYSTEM TYPE(S)	VRV Package Systems
CONTACT PERSON:	Motoi Yamaha

Objectives

The tool's objective is to evaluate performance of VRV packaged HVAC systems, whose capacity is almost impossible to measure when operating. The tool predicts energy consumption of the VRV systems from refrigeration cycle simulations without measuring their cooling capacity directly. By comparing predicted energy consumption with one that is actually measured malfunction or inadequate operation may be detected. Improved operations and management could result.

Functions

The functions of this tool are:

188

- energy consumption calculation for given building and weather data;
- detection of inadequate operation; and
- determination of base line from calculation in certain period.

Functionality is shown in the IDEFØ flow chart in Figure 0-17.

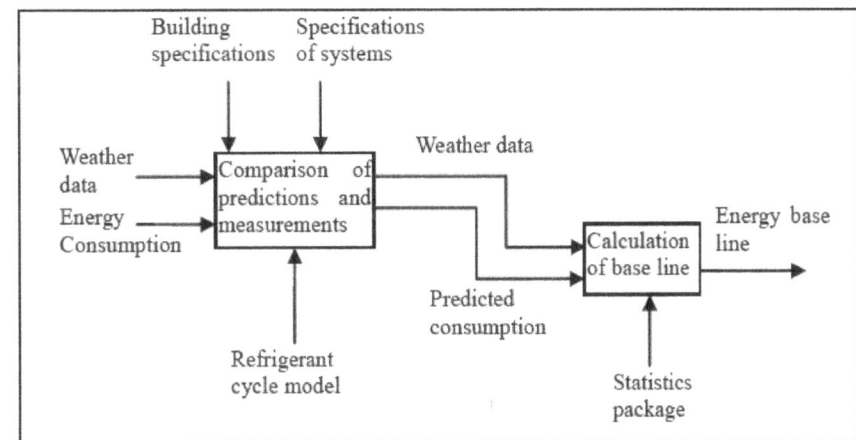

Figure 0-17 IDEFØ Flow of the tool

Continuous comparison between estimated and measured energy consumption can detect:

- Design errors (e.g., undersized or oversized equipment);
- Installation problems (e.g., excessive length of refrigerant pipes);
- Faults in system control; and
- Inappropriate room temperature settings where occupants control VRV systems individually.

Data Management

Input data are energy consumption of VRV systems, outdoor temperature, humidity and building and system configurations.

The analytical engine produces the comparison of estimation and measurements. Discrepancies reveal system faults or poor operation. Tools can be used for ongoing system analysis.

It is impossible to conduct load calculation under exact conditions used by the building. The tool should be calibrated for the specific building. This is under development.

Implementation

The tool was developed for university buildings. Since building management systems were not applied to such buildings, the tool is currently being used in off-line calculations. The tool combines computer program and package software; it could be implemented in any existing buildings whose data can be presented in a common file format.

Operability

Tool programs consist of:

- heat load calculation programs: independent from the tool itself and that use software such as EnergyPlus or TRNSYS.
- energy consumption predictions: a dedicated program that uses a simulated refrigerant cycle.
- data analysis: comparison and statistical analysis are made using Excel, R, etc.

Analytical Engine

Cooling and heating load are calculated from load calculation programs. The estimated energy consumption of VRV systems is obtained from the model that simulates the refrigeration cycle of VRV systems. The comparison will be presented.

End Users

The tool is not fully developed for building operators or energy manager unaccustomed to using computer programming. Therefore end users are currently limited to researchers.

Benefits

Although quantitative benefit was not obtained using the tool, it can evaluate performance of VRV systems, which is difficult measure.

Target Systems to be Tested

The tool is used exclusively with VRV packaged HVAC systems.

TOOL NAME	OH Saver
COUNTRY / ORGANIZATION	Japan / Hitachi Plant Technologies, LTD.
TARGETED BUILDING / SYSTEM TYPE(S)	non-residential (office)building / overall performance
CONTACT PERSON:	Hiroo Sakai

Objectives

The tool is aimed at monitoring the operating condition for HVAC system operation under the most adequate set-points to ensure minimal energy consumption. The tool is also used to optimize operation but for commissioning the whole system, sub-systems and equipments.

Functions

The main functions of the tool are to:

- monitor operating condition and current set-points;

- calculate optimum operating condition and set-points to minimize total energy consumption;

- calculate the difference between optimum operation and normal manual operation; and

- edit data to analyze performance trends for purposes of commissioning.

Data Management

Data management operations include:

- monitoring data every 5 minutes;

- filing data on temperature and flow rate for chilled water, condenser water, air supply/return;

- making a list of set-points from optimization calculation; and

- editing the operating data for commissioning.

Implementation

The tool is available for all office building HVAC system operation except non-central HVAC systems, such as a multi-packaged system. It requires data monitoring and directing devices and a data transferring system like Ethernet, BACnet and LONTALK.

Operability

The tool is developed for application in the automated operation of HVAC system. It requires an adjustable period of one season to adjust equipment objects (mathematical models to explain the operating characters) as they are on site. An IDEF representation of the tool is shown in Figure 0-18.

Analytical Engine

The analysis is driven by the following evaluation function:

$$f(x) = k_E \sum E_i(x) + k_G \sum G_i(x)$$, where:

$f(x)$: Evaluation function;
x: Optimization variable;
kE: Primary energy conversion coefficient for electricity[kJ/kWh];
kG: Primary energy conversion coefficient for gas [kJ/ m3];
$Ei(x)$: Consumed power [kW]; and
$Gi(x)$: Gas consumption volume [m3/h] .

End Users

Non-residential building owners

Benefits

In initial use in the Tokyo area, about 21 % of the energy used for HVAC systems was saved throughout the entire year and 25 % saved in summer.

Target Systems to be Tested

Non-residential buildings

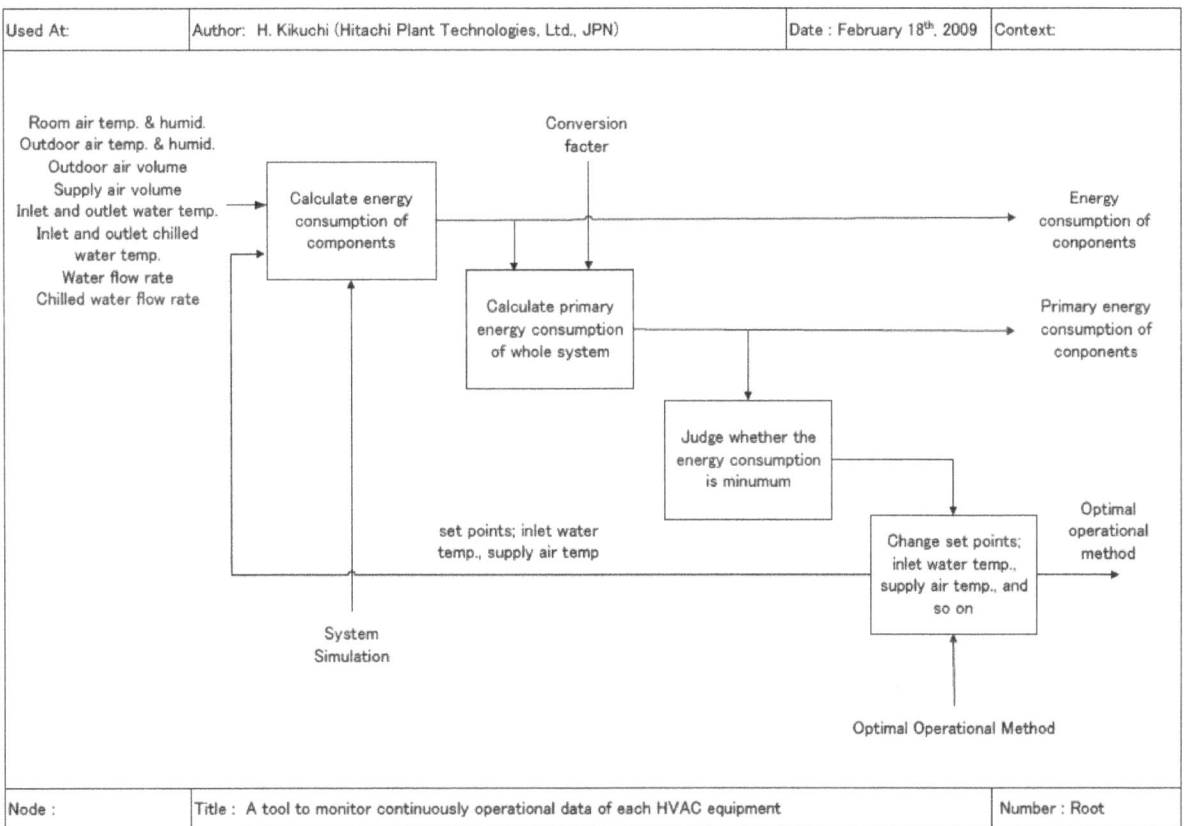

| Used At: | Author: H. Kikuchi (Hitachi Plant Technologies, Ltd., JPN) | Date : February 18th, 2009 | Context: |

Node : | Title : A tool to monitor continuously operational data of each HVAC equipment | Number : Root

Figure 0-18 IDEFØ Flow of the OH Saver tool

No. SUBTASK B: Commissioning and Optimization of Existing Buildings

TOOL NAME	Initial Cx Tool for HVAC System in Large Enclosure
COUNTRY / ORGANIZATION	Japan/SANKO Air Conditioning CO.,LTD.
TARGETED BUILDING / SYSTEM TYPE(S)	Any Type
CONTACT PERSON:	Mingjie Zheng

Objectives

The tool conducts initial commissioning at the design phase of large enclosures such as gymnasiums. The tool can be used to predict indoor climate and to detect possible design faults.

Functions

Figure 0-19 shows the IDEFØ diagram and tool function. Figure 0-20 is a detailed diagram of Activity B0 in Figure 1(1).

The tool's main functions are:

To obtain room air temperature and flow direction distribution by CFD. (Activity A0 in Figure 0-19)

- To calculate supply air temperature by PID simulation. (Activity B0 in Figure 1(1))

- To calculate pick-up time. (Activity C0 in Figure 1(1))

- With this tool, analysis can be performed to optimize design and operation parameters.

Data Management

The 10-minutes interval database is necessary to test and verify the simulation model. The input data include indoor air temperature gradient, supply air temperature and volume, building skin material characteristics, outside surface temperature of building skin, and control parameters of supply air temperature.

The output data include indoor air temperature gradient, the warm-up time at the peak-load day.

Training data is the same as input data. In the training process, the output data will be compared with measured values.

Implementation

This tool comprises a CFD software available at stores, a PID analysis tool developed by the author, and a data transfer text file.

Operability

A convenient interface comes installed with the CFD software. Because the PID analysis tool is easy to use, there is no need to develop another interface.

Analytical Engine

CFD, PID analysis tool.

End Users

Commissioning authority, system designer.

Benefit

This tool is very useful in detecting design errors in HVAC systems used in large enclosures.

Target Systems to be Tested

Two case studies will be conducted in gymnasiums in Japan.

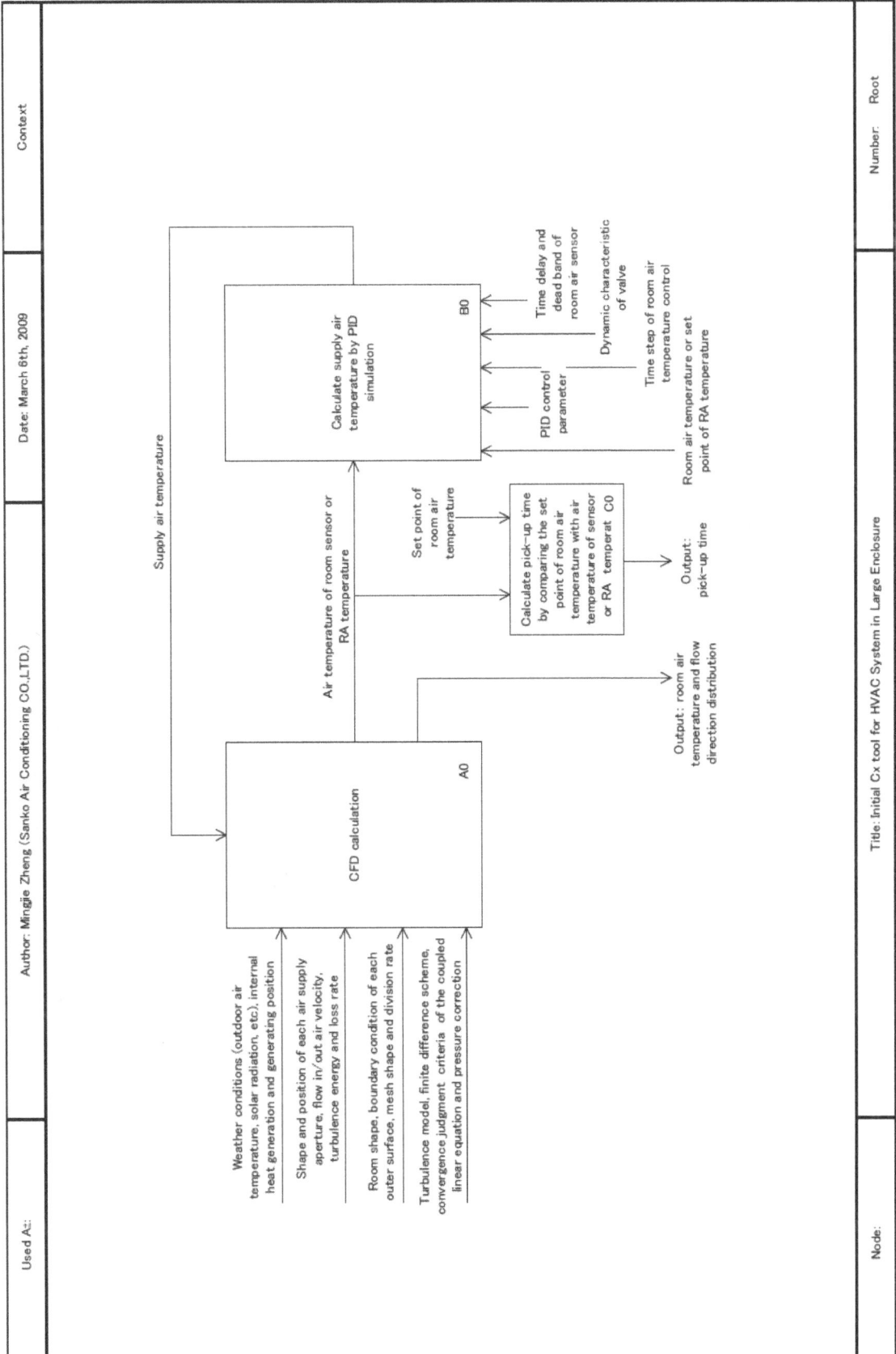

Supply air temperature

Calculate supply air temperature by PID simulation

B0

Time delay and dead band of room air sensor

Dynamic characteristic of valve

Time step of room air temperature control

PID control parameter

Room air temperature or set point of RA temperature

Air temperature of room sensor or RA temperature

Set point of room air temperature

Calculate pick-up time by comparing the set point of room air temperature with air temperature of sensor or RA temperat   C0

Output: pick-up time

CFD calculation

A0

Output: room air temperature and flow direction distribution

Weather conditions (outdoor air temperature, solar radiation, etc.), internal heat generation and generating position

Shape and position of each air supply aperture, flow in/out air velocity, turbulence energy and loss rate

Room shape, boundary condition of each outer surface, mesh shape and division rate

Turbulence model, finite difference scheme, convergence judgment criteria of the coupled linear equation and pressure correction

Figure 0-19 IDEF0 tool diagram (Root)

195

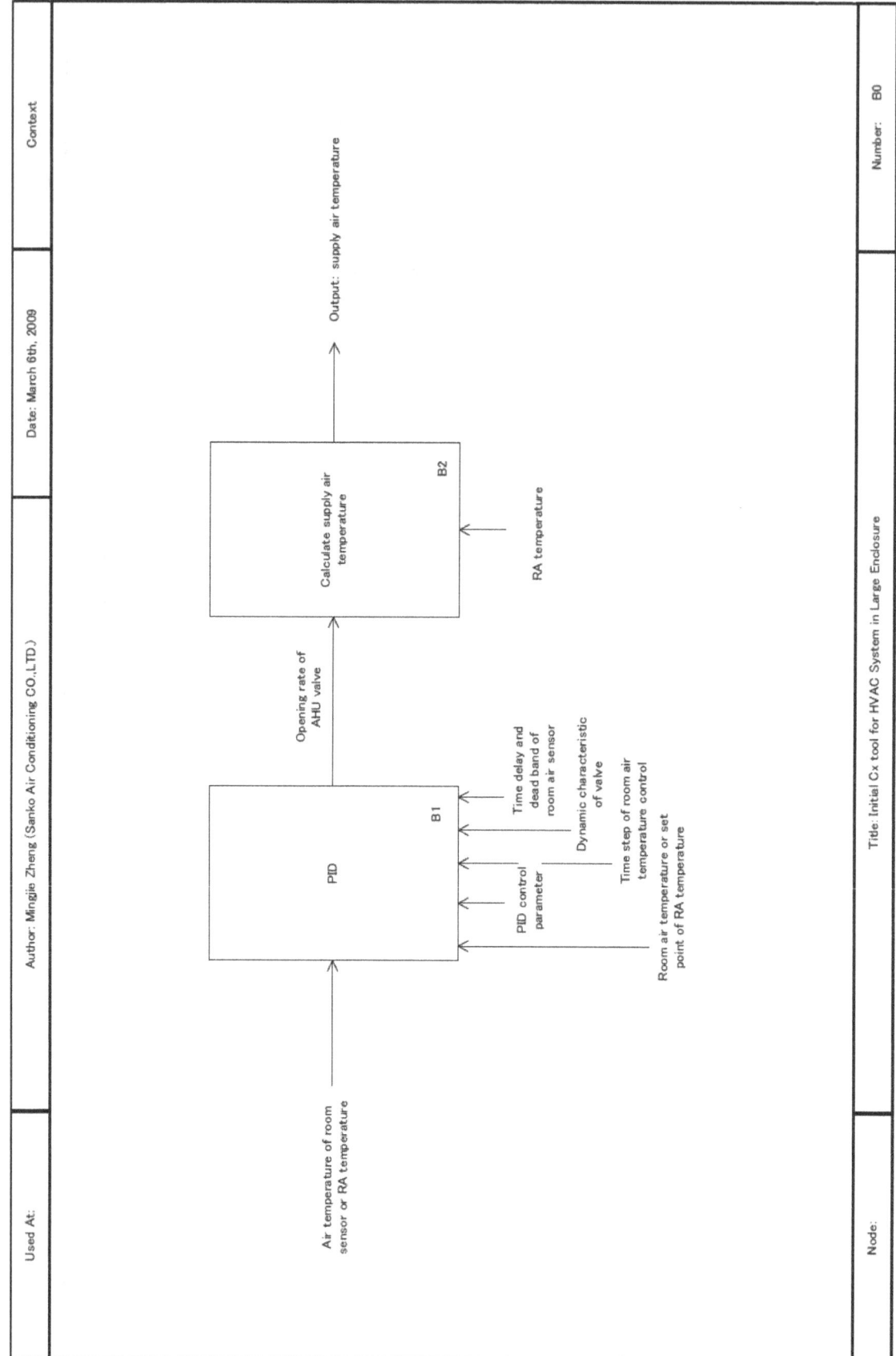

Air temperature of room sensor or RA temperature

PID

B1

PID control parameter

Time step of room air temperature control

Dynamic characteristic of valve

Time delay and dead band of room air sensor

Opening rate of AHU valve

Calculate supply air temperature

B2

RA temperature

Output: supply air temperature

Room air temperature or set point of RA temperature

Figure 0-20 IDEF0 tool diagram (B0)

TOOL NAME	Ongoing Type FDD Tool with Pattern Recognition for VAV systems
COUNTRY / ORGANIZATION	Japan/SANKO Air Conditioning CO.,LTD.
TARGETED BUILDING / SYSTEM TYPE(S)	Any Type
CONTACT PERSON:	Mingjie Zheng, Song Pan

Objectives

The aim of this tool is to detect and diagnose existing faults in HVAC systems at the operational phase as an aspect of ongoing commissioning.

Functions

Figure 0-21 shows the IDEFØ diagram and tool functions. Figure 0-22 and Figure 0-23 are detailed diagrams of Activity B0 and C0 in Figure 1(1) respectively.

The tool's main functions are:

- To Perform Fourier analysis for measurement. (Activity A0 in Figure 1(1))

- To calculate detection and diagnosis parameter with measurement and their Fourier transformation by statistical method. (Activity B0 in Figure 1(1))

- To select optimum detection and diagnosis vector by detection rate increment method. (Activity C0 in Figure 1(1))

- To judge system state using optimum detection and diagnosis figure or Mahananobis's generalized distance.(Activity D0 and E0 in Figure 1(1))

With this tool, parameters expressing the performance and characteristics of VAV systems can be calculated with measured data and their Fourier transformation value. Based on these parameters, the system state can be judged as being normal to faulty.

Data Management

A 10 min intervals database is necessary to calibrate the calculating model. The data consist of air temperature and supply air volume of each VAV unit on VAV HVCA system, in normal state and various faulty operations.

Create data base in normal state: After modulating the VAV HVAC system well and ensuring it is in operating normally, the air temperature and supply air volume of each VAV unit are measured at 10 min intervals to form the normal state database.

Create data base in various faulty states: Run dynamic simulation program to calculate air temperature and supply air volume of each VAV unit in a normal state. After ensuring simulated results coincide with the normal database gathered above, run the program again in various faulty states at 10 min intervals to form the faulty data base.

Wherever possible, use data from existing BEMS for normal state and various faulty state operations.

The optimal parameter vector calculated using the air temperature and air volume data of the VAV unit to be measured online is then evaluated using Mahananobis' generalized distance for normal or faulty operation.

Implementation

This tool is developed with FORTRAN90.

Operability

A user interface will be developed to continuously monitor input data of the HVAC system.

Analytical Engine

A simulation tool was developed by authors using FORTRAN90.

End Users

Commissioning authority, system operator.

Benefit

This tool is useful in detecting existing faults in HVAC systems and diagnosing cause.

Target Systems to be Tested

The tool has been used in case studies conducted on both VAV systems.

Used At:

Author: Mingjie Zheng and Song Pan (Sanko Air Conditioning CO.,LTD.)

Date: March 6th, 2009

Context

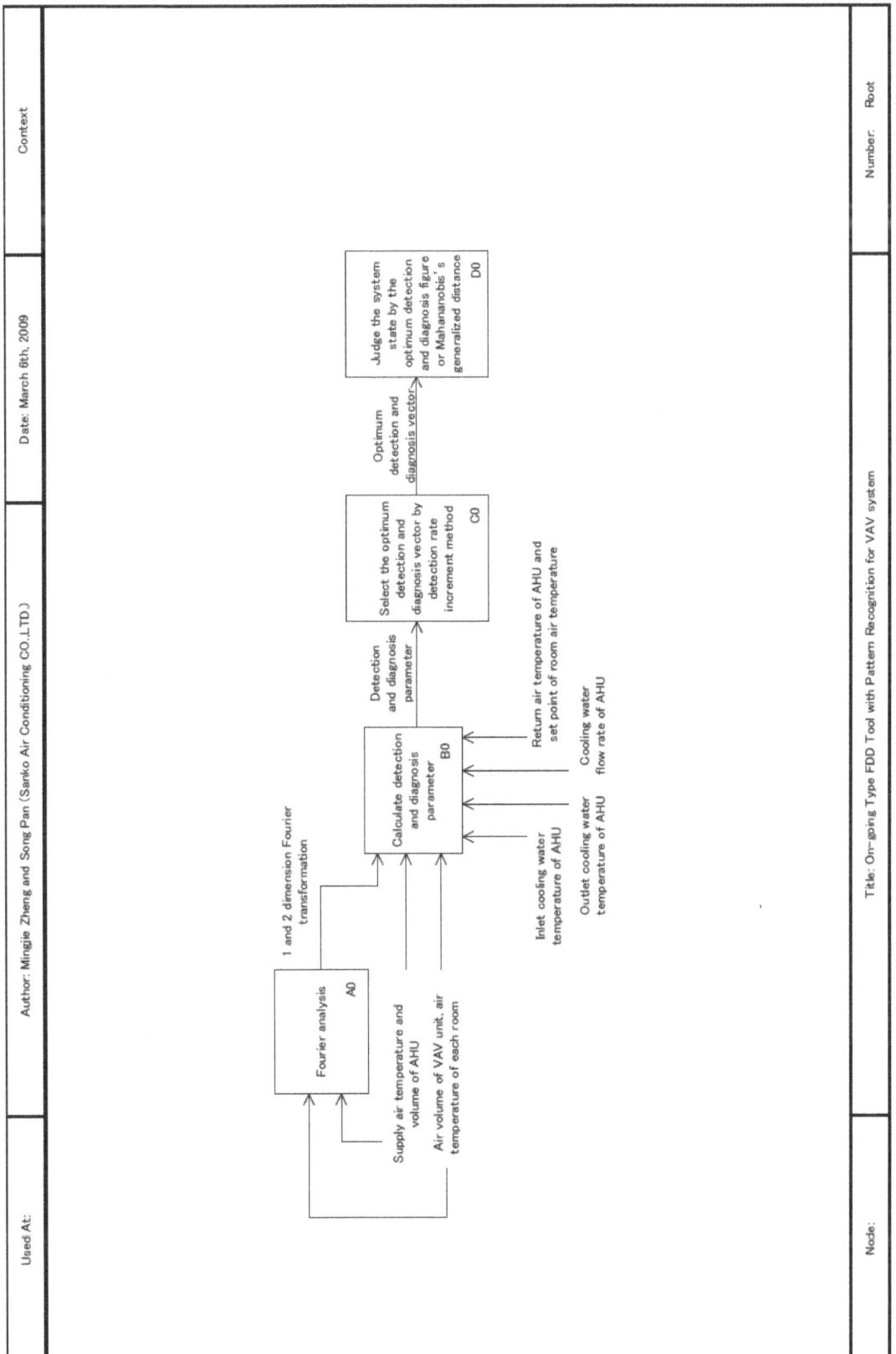

Figure 0-21 IDEF0 tool diagram (Root)

199

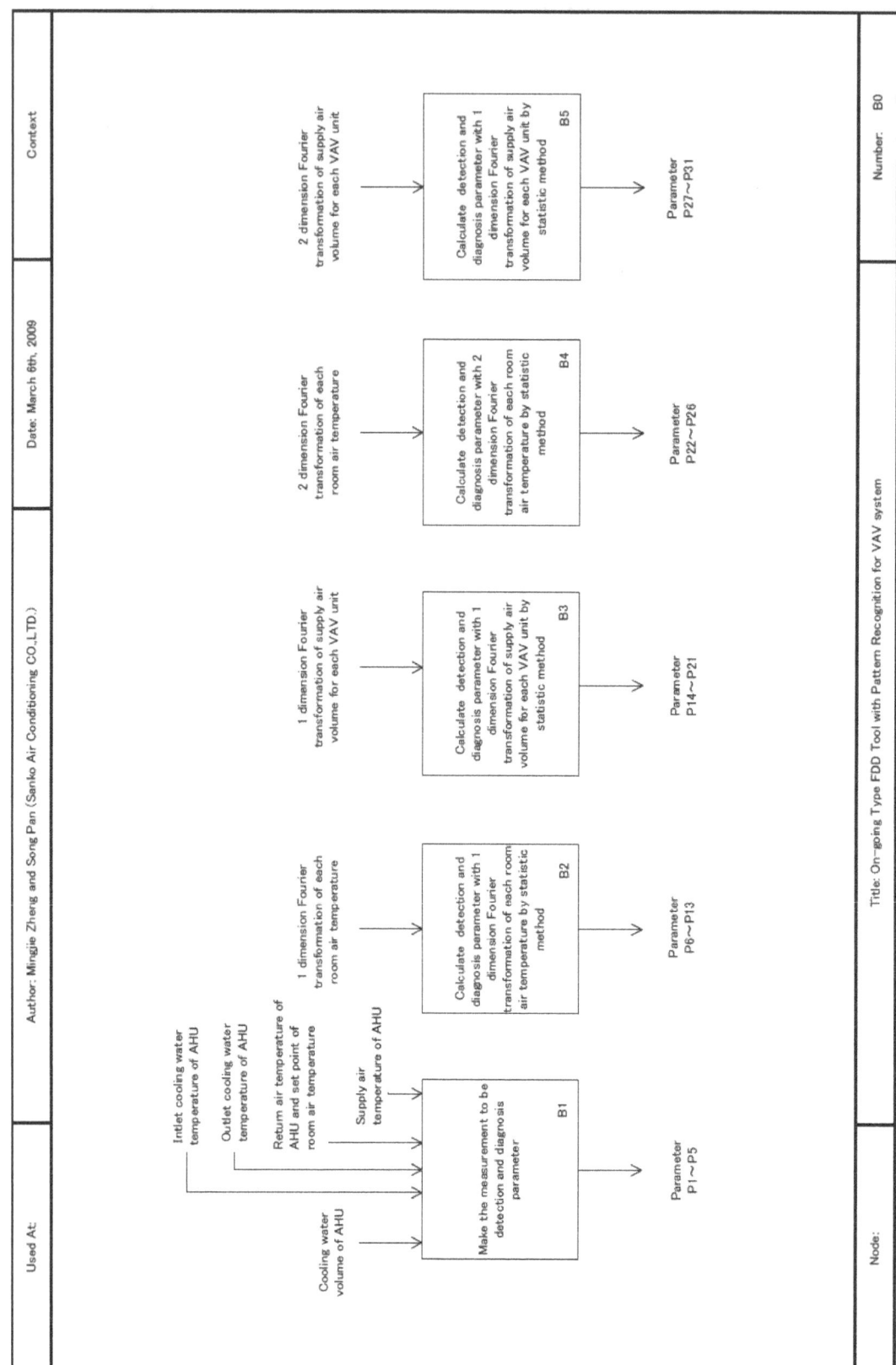

2 dimension Fourier transformation of supply air volume for each VAV unit

Calculate detection and diagnosis parameter with 1 dimension Fourier transformation of supply air volume for each VAV unit by statistic method

B5

Parameter P27~P31

2 dimension Fourier transformation of each room air temperature

Calculate detection and diagnosis parameter with 2 dimension Fourier transformation of each room air temperature by statistic method

B4

Parameter P22~P26

1 dimension Fourier transformation of supply air volume for each VAV unit

Calculate detection and diagnosis parameter with 1 dimension Fourier transformation of supply air volume for each VAV unit by statistic method

B3

Parameter P14~P21

1 dimension Fourier transformation of each room air temperature

Calculate detection and diagnosis parameter with 1 dimension Fourier transformation of each room air temperature by statistic method

B2

Parameter P6~P13

Inlet cooling water temperature of AHU

Outlet cooling water temperature of AHU

Return air temperature of AHU and set point of room air temperature

Supply air temperature of AHU

Cooling water volume of AHU

Make the measurement to be detection and diagnosis parameter

B1

Parameter P1~P5

Figure 0-22 IDEF0 tool diagram (B0)

200

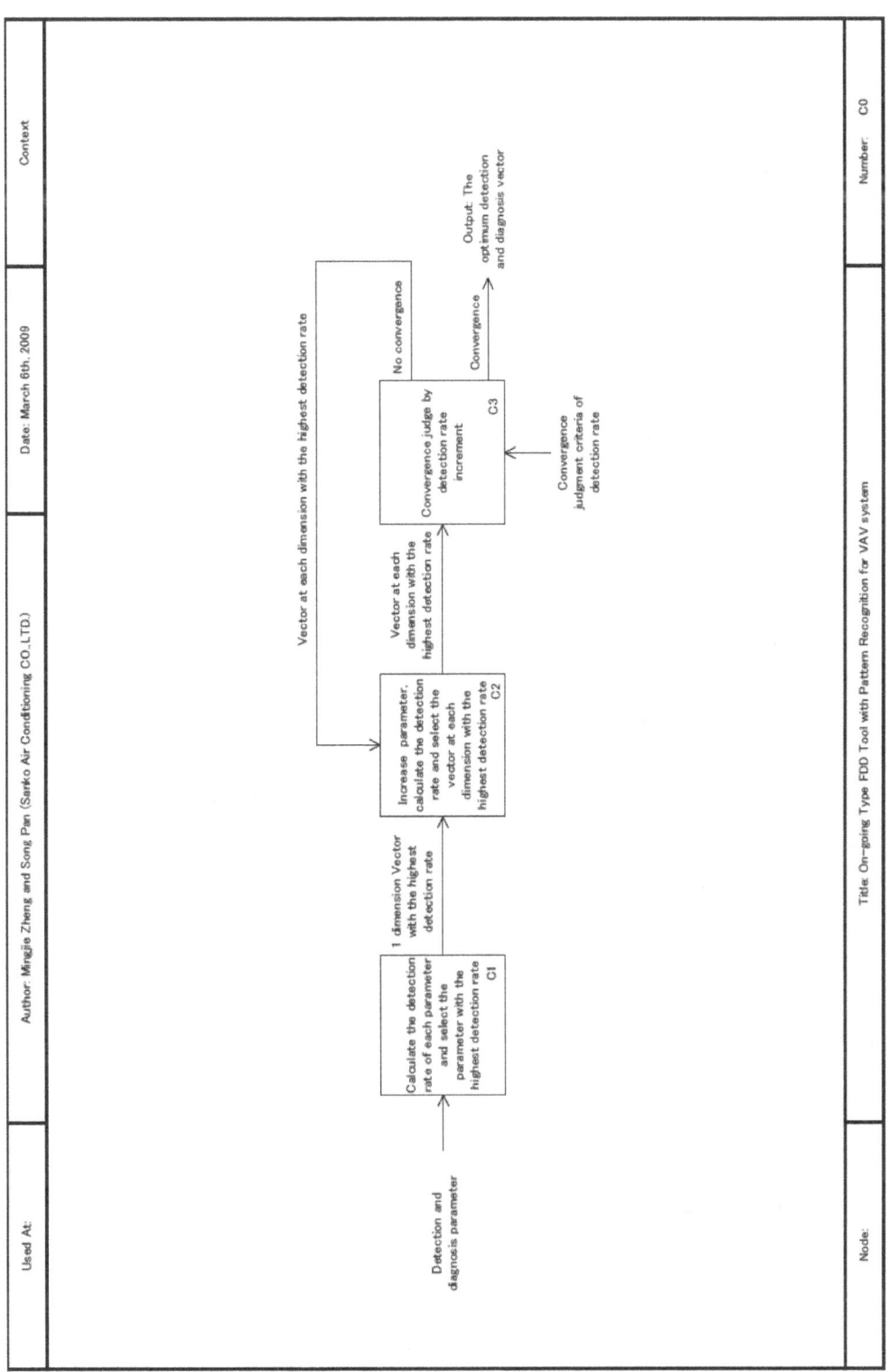

Figure 0-23 IDEF0 tool diagram (C0)

201

TOOL NAME	A Tool for Optimizing Heating/Cooling Plant Operation
COUNTRY / ORGANIZATION	Japan / Kyoto University
TARGETED BUILDING / SYSTEM TYPE(S)	Any type
CONTACT PERSON:	Harunori Yoshida

Objectives

This tool checks current operation of a heating/cooling plant visually and optimizes operation automatically. The tool consists of two modules: a visualization module and a performance simulation module. The visualization module can describe current operation in carpet graphics and detect improper operation. The performance simulation module consists of equipment and equipment control models. This module aims at finding an optimized operation method through simulating the performance of the plant running at different levels.

Functions

The tool consists of two modules that offer two different functions:

- Operation visualization module. It uses carpet graphics to describe the current situation, thereby easily detecting improper operation.
- Performance simulation module. This module consists of the equipment models configured in a heating/cooling plant and the models of equipment control. With it, the performance of a proposal for optimizing a heating/cooling plant operation can be tested.

Data Management

The following data are used:

- Input data for operation visualization module: average water temperature of thermal storage tank, time to determine thermal storage mode (storage or discharge), power consumption of heat source machines
- Input data for Performance simulation module: weather data, water flow rate, water temperature set point, heating/cooling load, control set points
- Output of operation visualization module: operation visualization carpet graphics
- Output of performance simulation module: plant energy consumption and operation optimization proposal

Figure 0-24 shows inputs and output of the operation visualization module.

202

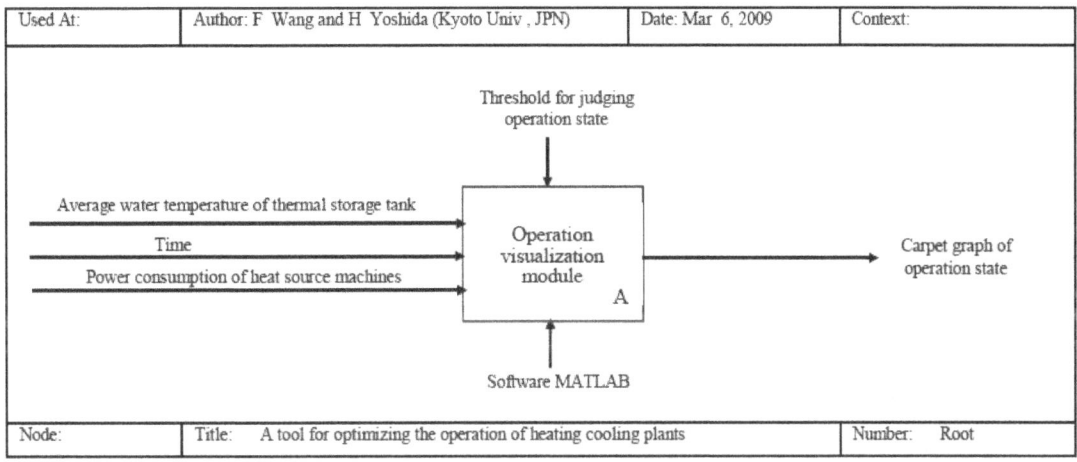

| Used At: | Author: F Wang and H Yoshida (Kyoto Univ , JPN) | Date: Mar 6, 2009 | Context: |

Threshold for judging operation state

Average water temperature of thermal storage tank

Time

Power consumption of heat source machines

Operation visualization module A

Carpet graph of operation state

Software MATLAB

| Node: | Title: A tool for optimizing the operation of heating cooling plants | Number: Root |

Figure 0-24 IDEFO diagram of the Operation visualization module (Root)

Figures 0-25, 0-26, and 0-27 show the inputs, output and internal data flow of the Performance simulation module.

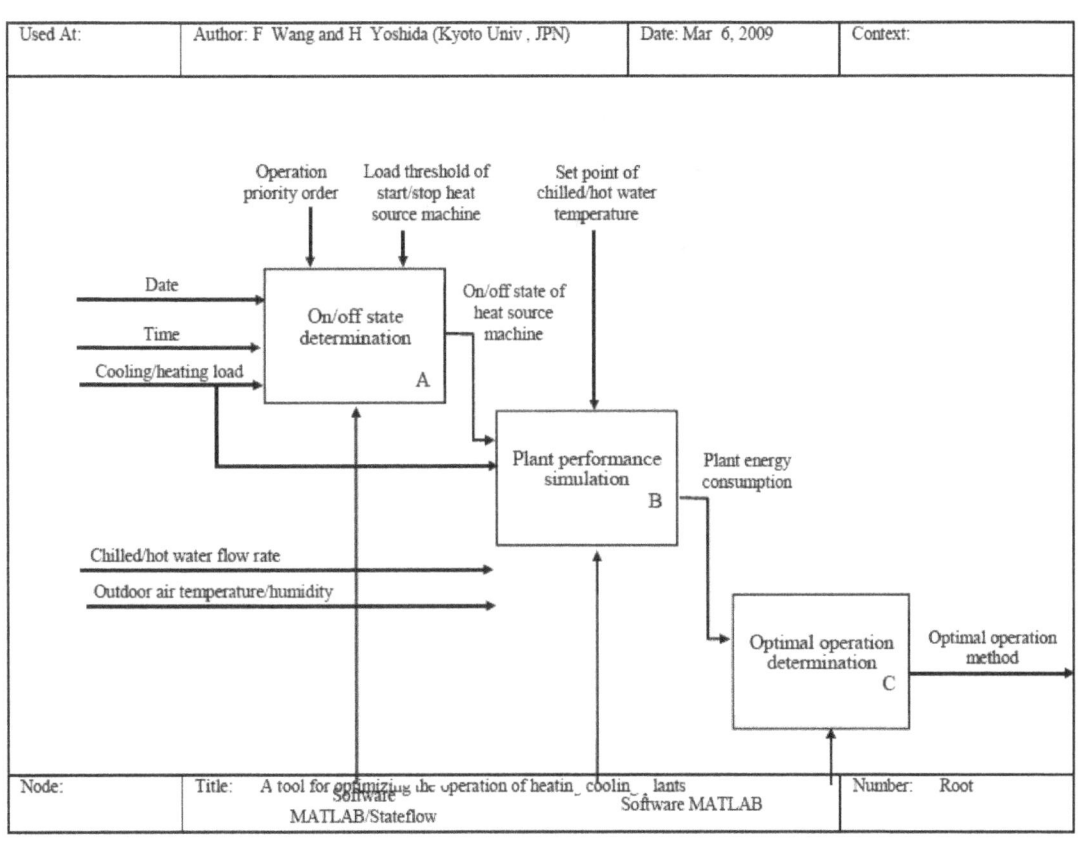

| Used At: | Author: F Wang and H Yoshida (Kyoto Univ , JPN) | Date: Mar 6, 2009 | Context: |

Operation priority order

Load threshold of start/stop heat source machine

Set point of chilled/hot water temperature

Date

Time

Cooling/heating load

On/off state determination A

On/off state of heat source machine

Plant performance simulation B

Plant energy consumption

Chilled/hot water flow rate

Outdoor air temperature/humidity

Optimal operation determination C

Optimal operation method

| Node: | Title: A tool for optimizing the operation of heating cooling plants | Number: Root |

Software MATLAB/Stateflow

Software MATLAB

Figur

e 0-25 IDEFO diagram of performance simulation module (Root)

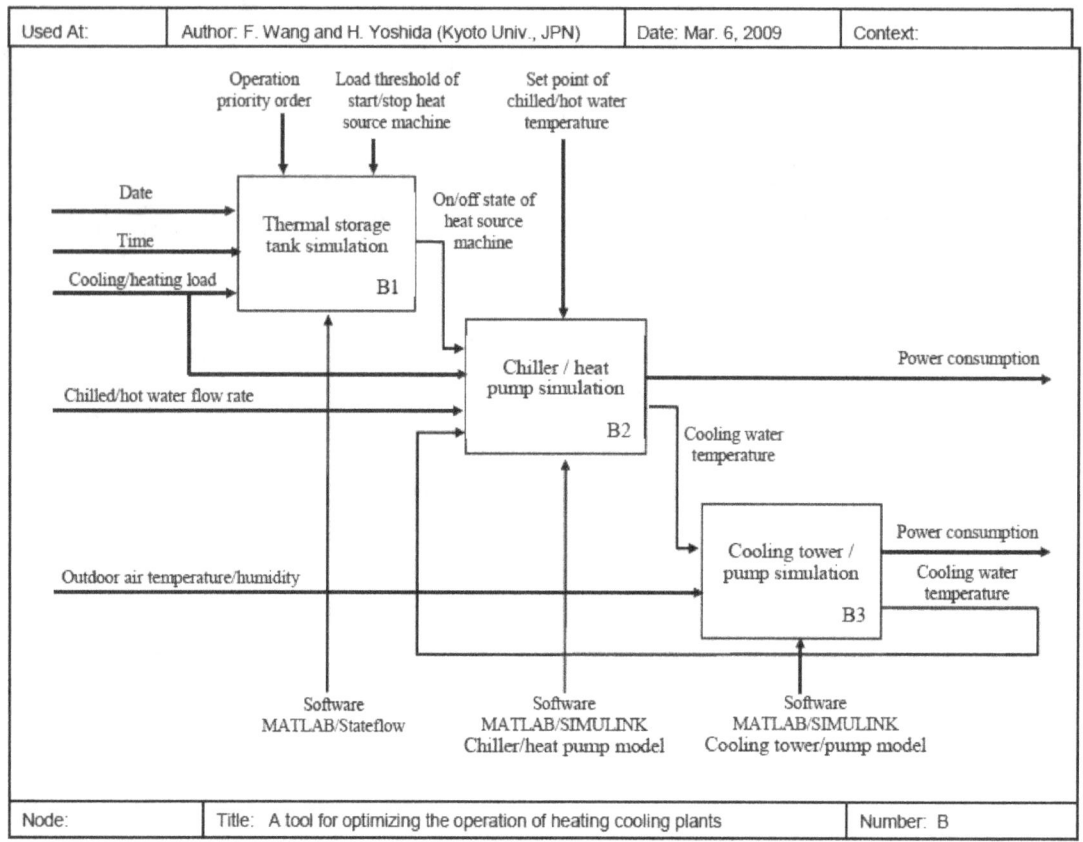

Operation priority order

Load threshold of start/stop heat source machine

Set point of chilled/hot water temperature

Date

Time

Cooling/heating load

Thermal storage tank simulation

B1

On/off state of heat source machine

Chiller / heat pump simulation

B2

Power consumption

Chilled/hot water flow rate

Cooling water temperature

Cooling tower / pump simulation

B3

Power consumption

Cooling water temperature

Outdoor air temperature/humidity

Software MATLAB/Stateflow

Software MATLAB/SIMULINK Chiller/heat pump model

Software MATLAB/SIMULINK Cooling tower/pump model

| Node: | Title: A tool for optimizing the operation of heating cooling plants | Number: B |

Figure 0-26 IDEFØ diagram of performance simulation module (B)

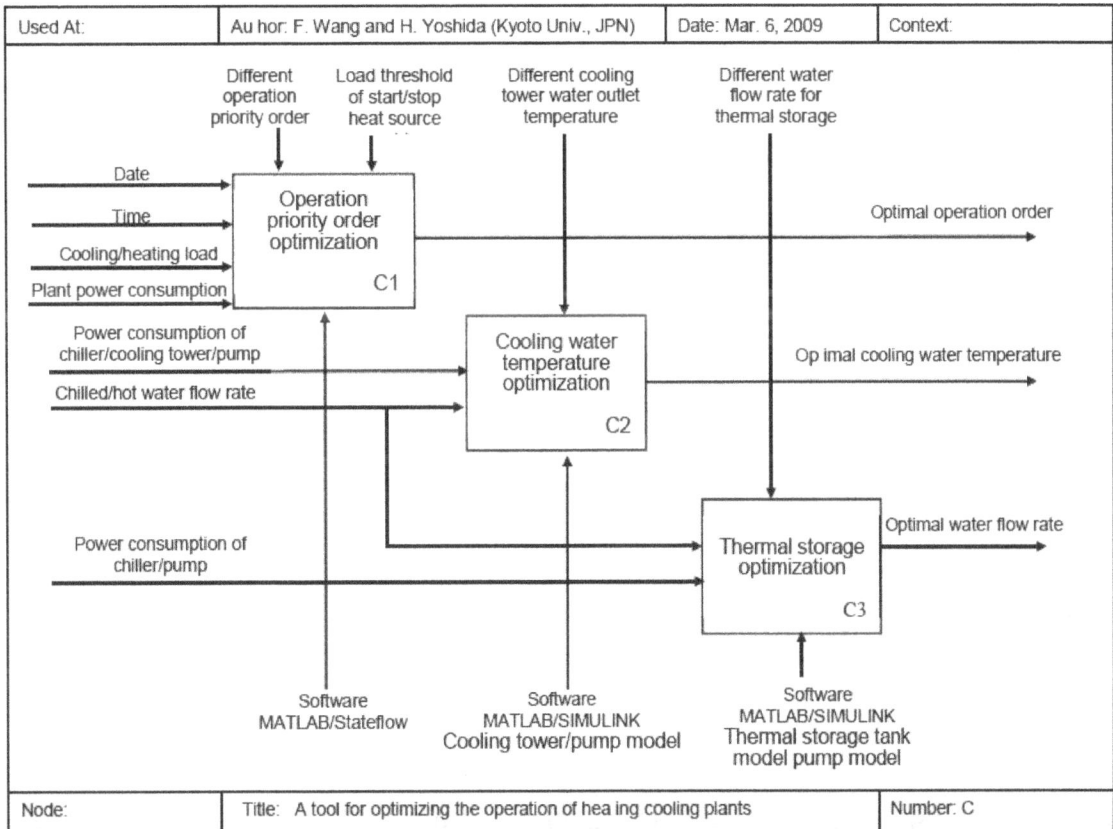

| Used At: | Au hor: F. Wang and H. Yoshida (Kyoto Univ., JPN) | Date: Mar. 6, 2009 | Context: |

Operation priority order optimization — C1

Cooling water temperature optimization — C2

Thermal storage optimization — C3

| Node: | Title: A tool for optimizing the operation of hea ing cooling plants | Number: C |

Figure 0-27 IDEFØ diagram of performance simulation module (C)

Implementation

The software is developed in MATLAB language and MATLAB Simulink and Stateflow toolbox is used to develop the tool. The tool can be implemented in a MATLAB environment or as a stand-alone.

Operability

To make the tool easy to use, a graphic user interface is developed using the MATLAB Simulink and Stateflow function.

Analytical Engine

The MATLAB curve fitting function is used to develop equipment models and their performance curves. Operation mode (running combined equipment) is determined automatically and visually using Stateflow function. All equipment is connected using the Simulink function to develop a whole plant model and simulate its energy performance.

End users can be:

- Heating/cooling plant designer;
- Heating/cooling plant operator; and/or
- Heating/cooling plant commissioning authority.

Benefits

Visualizes current operation using carpet graphics for easy detection of improper operation.
Automates and visualizes operation mode determination to save manpower and time.
Finds optimal operation method quickly, which is impossible through experimentation

Target Systems to be Tested

The tool is tested in two district heating cooling plants in Osaka Japan. Proper plant operation is confirmed visually. Performances of recommended improvements are checked. Experiments are also conducted to verify recommendations.

No. 12   SUBTASK B: Commissioning and Optimization of Existing Buildings

TOOL NAME	Tool for Optimizing HVAC System with Ground Thermal Storage
COUNTRY / ORGANIZATION	Japan / Kyoto University
TARGETED BUILDING / SYSTEM TYPE(S)	HVAC system with Ground Thermal Storage
CONTACT PERSON:	Harunori Yoshida

Objectives

The tool can simulate the performance of HVAC systems with ground thermal storage systems. Water cooled by cooling towers is circulated through pipes pre-installed in foundation piles to cool the ground in winter; in summer the cool water is fed to AHU cooling coils to pre-cool supply air. Although the system utilizes natural energy, if operated improperly the system may consume more energy than a common HVAC system with chillers, due to the large amount of energy used for water circulation. The tool can determine the system's operational method.

Functions

Figure 0-28. IDEF0 diagram of the tool's functions. Figures 0-29, 0-30, 0-31, and 0-32 are expanded diagrams of Activity A in Figure 0-28, Activity A2 in Figure 0-29 Activity A3 in Figure 0-30 and Activity B in Figure 0-28 respectively.

The tool's main functions are:

- To calculate energy consumption of the HVAC system based on weather data, room air temperature, operation schedule and given operational methods. (Activity A in Figure 0-28)

- To calculate outlet water temperature from the pipes buried in the ground based on water temperature and water flow rate. (Activity B in Figure 0-28)

- To determine optimal operation values among the operation methods given. (Activity C in Figure 0-28)

Details of the functions and calculations are described in the expanded IDEF0 diagrams in Figures 0-28 to 0-32.

Data Management

Figure 0-28 shows the required information for the tool and its internal data flow. The input data are weather data, room air temperature, operation schedule, and operation methods. The output data are the optimal operation values. The measured time-series data are stored as a CSV file, which has a defined time label in the first 7 rows. These rows indicate year, month, day, day of the week, hour, minute and second, Rows 1 to Row 7.

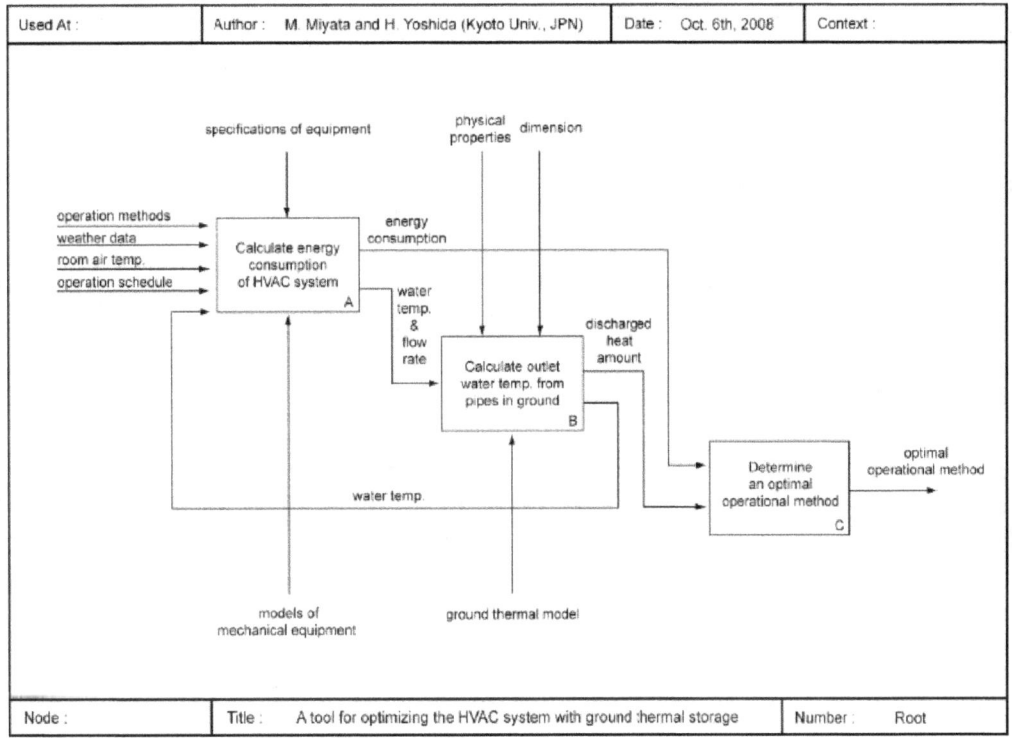

Figure 0-28 IDEF0 tool diagram(Root)

207

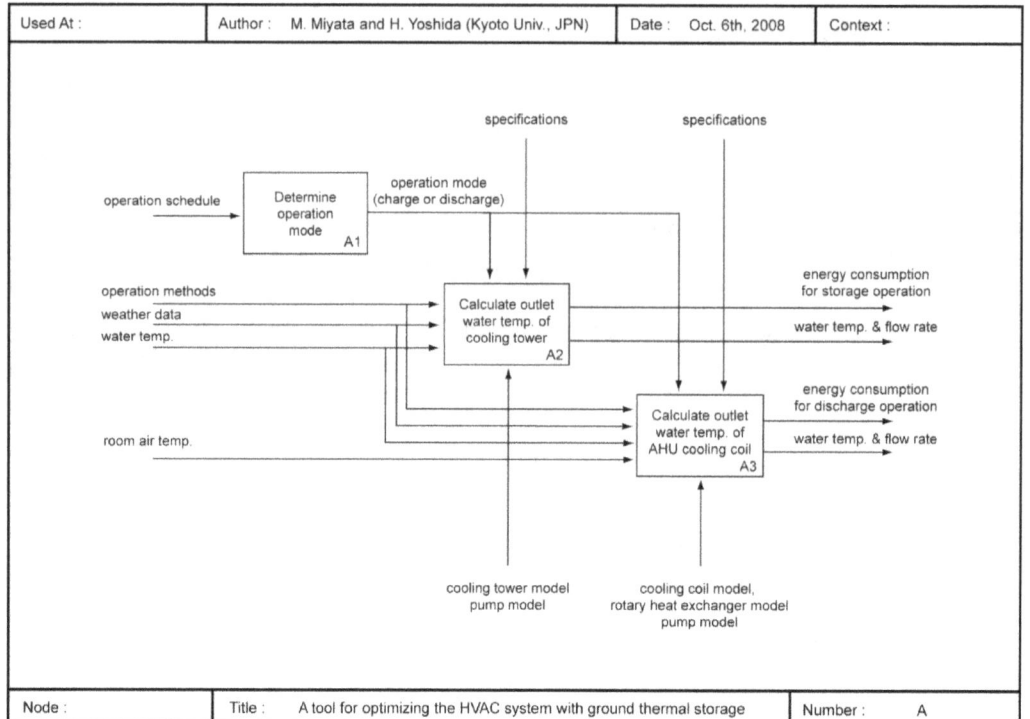

Figure 0-29 IDEFØ tool diagram (A)

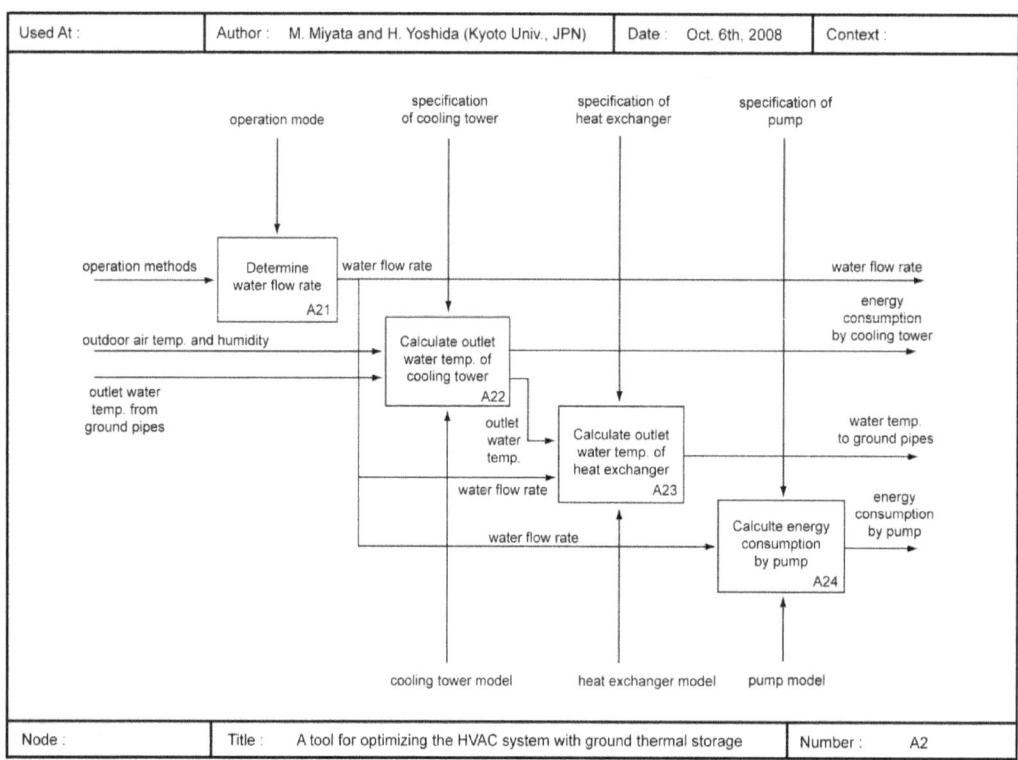

Figure 0-30 IDEFØ tool diagram (A2)

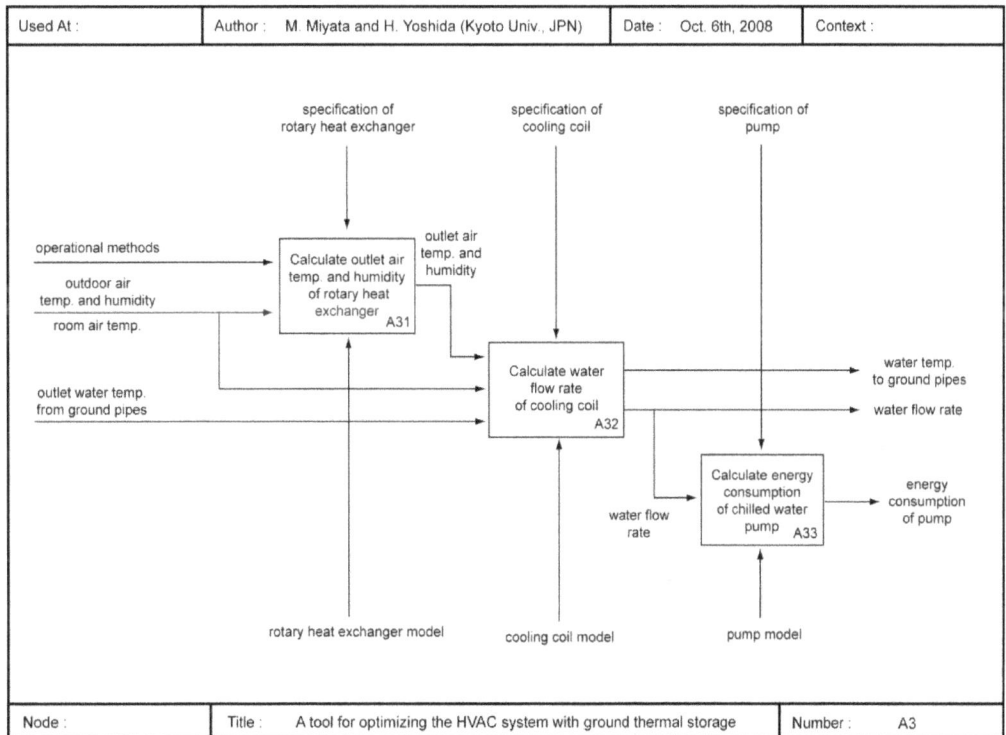

Figure 0-31 IDEFØ tool diagram (A3)

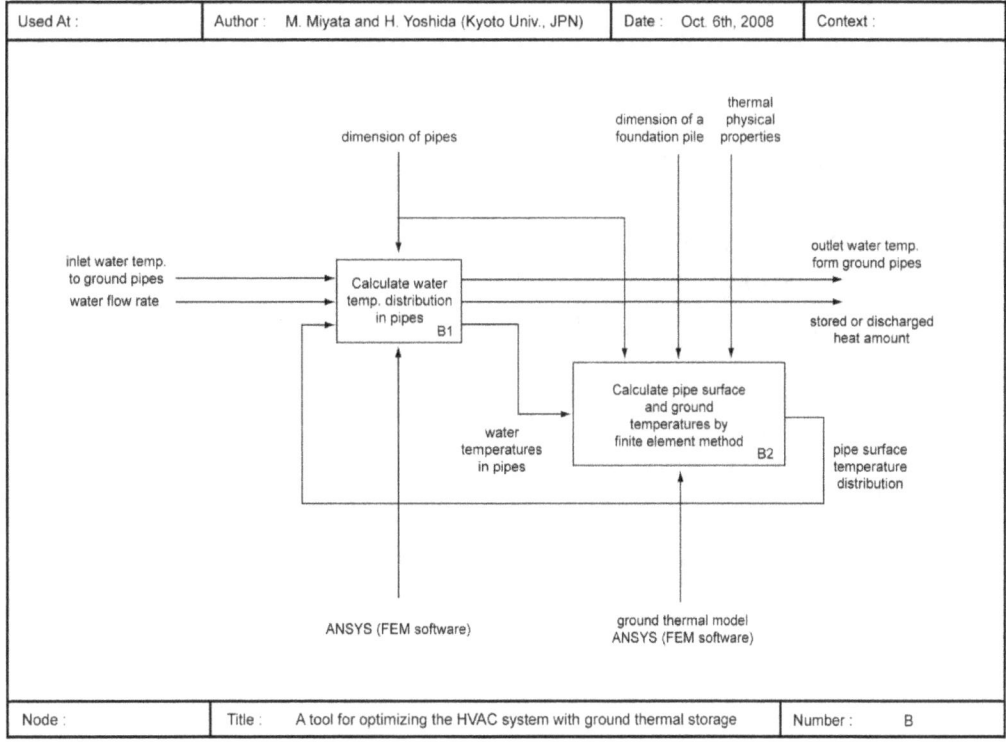

Figure 0-32 IDEFØ tool diagram (B)

209

Implementation

This tool runs on the MATLAB/Simulink platform and runs FEM thermal analysis software ANSYS in the background. .

Operability

We developed a graphical user interface using Simulink. The HVAC component model is expressed as a block. By double-clicking the block, a pop-up dialog box appears, enabling model parameters to be easily modified.

Analytical Engine

MATLAB, Simulink, ANSYS (FEM thermal analysis software)

End Users

Cx providers, Design engineers, Building operators

Benefits

The long heat transfer time underground makes it difficult to attempt various operational methods experimentally. Simulation therefore becomes a powerful system tool.

Target Systems to be Tested

The tool was used at an office building in Takamatsu, Japan. The building exchanges heat underground using foundation piles. In winter, circulated water chilled by cooling towers flows through tubes in the foundation piles to cool down in ground. In summer, circulated water cooled by the ground flows through cooling coils to cool down supply air.

No. 13   SUBTASK B: Commissioning and Optimization of Existing Buildings

TOOL NAME	A Simulation Tool to Estimate Baseline Energy
COUNTRY / ORGANIZATION	Japan / Kyoto University
TARGETED BUILDING / SYSTEM TYPE(S)	Any type
CONTACT PERSON:	Harunori Yoshida

Objectives

The tool can estimate energy baselines for ongoing commissioning or implementation of an energy saving retrofit, and make adjustments using operation conditions following the retrofit.

Functions

The tool consists of two modules: a heating/cooling load calculation module and an energy consumption estimation module, which consists of sub-models of the building's energy consuming equipment. Hourly energy consumption is calculated using estimated heating/cooling loads.

Data Management

The measured time-series data is stored in CSV files, which are assigned defined time labels in the first 7 rows. Rows 1 to 7 indicate year, month, day, day of the week, hour, minute and second, respectively.

Tool input is weather data, room air temperature and humidity, fresh intake air volume, air-conditioning system operation schedules, lighting systems and occupancy. Output is energy baseline adjusted by conditions following commissioning or an energy savings retrofit.

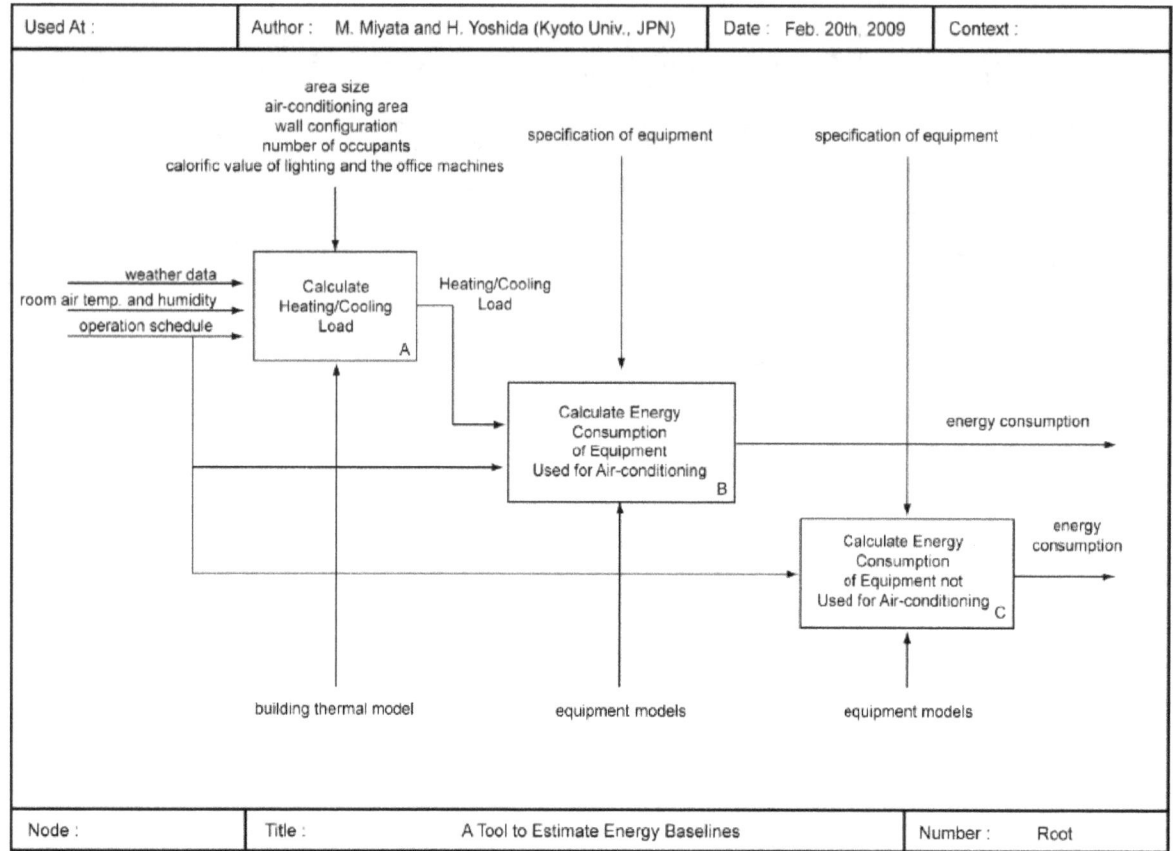

Used At :	Author : M. Miyata and H. Yoshida (Kyoto Univ., JPN)	Date : Feb. 20th. 2009	Context :

| Node : | Title : A Tool to Estimate Energy Baselines | Number : Root |

Figure 0-33 IDEFØ tool diagram (Root)

Implementation

This tool is developed on the MATLAB/Simulink platform.

Operability

We developed a graphical user interface using Simulink. By double-clicking the block, a pop-up dialog box appears, enabling model parameters to be easily modified.

Analytical Engine

Simulation based air-conditioning load calculation and building system component models by MATLAB/ Simulink

End Users

Cx providers, Energy Service Companies, Building operators

212

The tool can calculate a building's energy baseline using various conditions such as weather, occupancy and equipment operations.

Target Systems to be Tested

The case study was conducted at a mid-scale office building in Osaka Japan. The tool was used in the building; the model was tested using measured operational data.

No. 14 SUBTASK B: Commissioning and Optimization of Existing Buildings

TOOL NAME	Life-Cycle Energy Management (LCEM)
COUNTRY / ORGANIZATION	Japan / Government Building, Ministry of Land, Infrastructure and Transport (MLIT)
TARGETED BUILDING / SYSTEM TYPE(S)	Any
CONTACT PERSON:	TBD

Objectives

LCEM : Life-Cycle Energy Management is being developed to assess and evaluate HVAC energy performance throughout a building's lifecycle (e.g., planning, design, construction, operation, maintenance, refurbishment and demolition). The LCEM project was conducted for 3 years beginning in April 2004, under the leadership of the Japanese Ministry of Land, Infrastructure and Transport. As owner and property manager, the ministry's Government Buildings Department made initiatives for the development of LCEM .

The tool aims to provide a platform for shared measurement among all the various stakeholders concerned with each stage of the building's life cycle. It also aims to be operable in spreadsheets (e.g., MS Excel) and act as a visual aid to support communication.

Functions

The functions of this tool are :

- Basic static analysis by simulation – assessing energy performance for overall system, subsystems and/or individual equipment units and components being operated with different characteristics;
- Assessment to confirm performance at 1. equipment fabrication stage and 2. test operations/adjustment stage using design sate reference requirements;
- Assess performance by comparing design values and test operation status values (actual performance) during acceptance procedures and test operation, etc.; and

- Predict performance at off-peak times (not just peak load performance) under various outdoor conditions such as mid-summer, mid-winter and partial load seasons (rainy season and early winter).

Data Management

Input data:

Outdoor air condition/heat load in each room (hourly, all year) Users select load patterns from existing heat load estimation tools. Simplified calculation tool in spreadsheet is also being developed
Characteristic curve of each HVAC component (load-efficiency curve)

Output:

Status and energy consumption of each component
Suggestions for equipment size and division of capacity in terms of energy performance

Implementation

LCEM is being developed by Microsoft Excel $^{(TM)}$ with visual aid objects to support stakeholder communication.

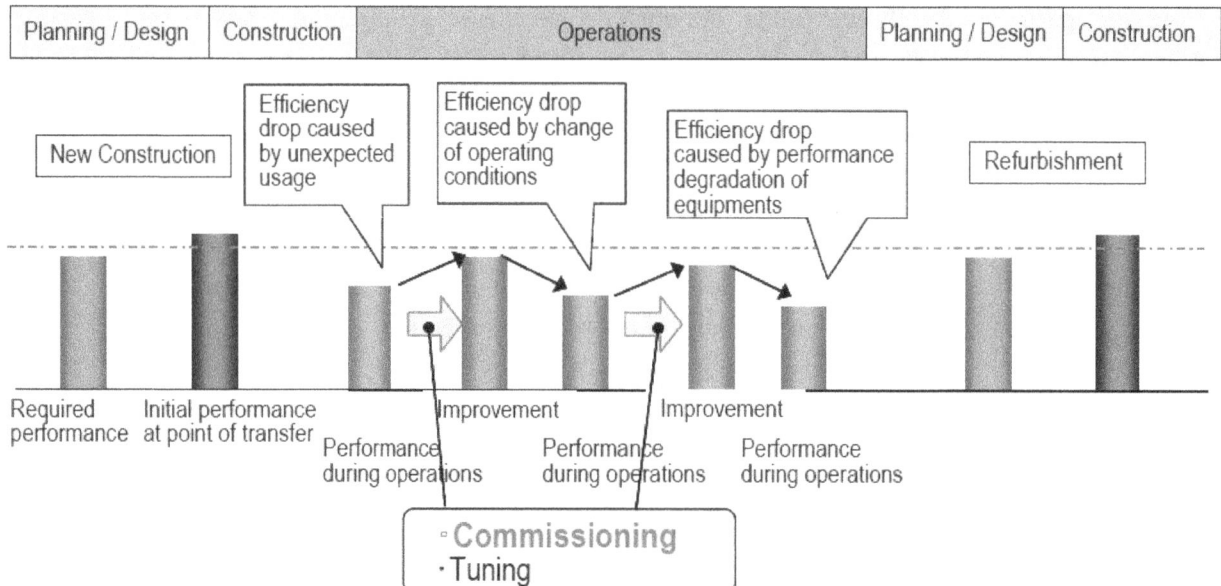

Figure 0-34 Life-cycle performance assessment and commissioning for HVAC systems

1. Each equipment unit is defined in a spreadsheet cell. A block of cells includes characteristics and specifications for one equipment unit. The block of cells is named an "object composed cells" or just object.

2. Users select, cut and paste an object from the list in order, from left to right, to make an HVAC model.

3. Users input control parameters (e.g., cooling water temperature).

4. Cross references occur almost every time with any change to data. This prompts iteration calculation in the worksheet.

| Outside air WB (Boundary) | Cooling Tower Object | Cooling Water Pump object | Gas Absorption Chiller Objects | Primary Pump Object | Water Flow rate Return Water Temp. (Boundary) |

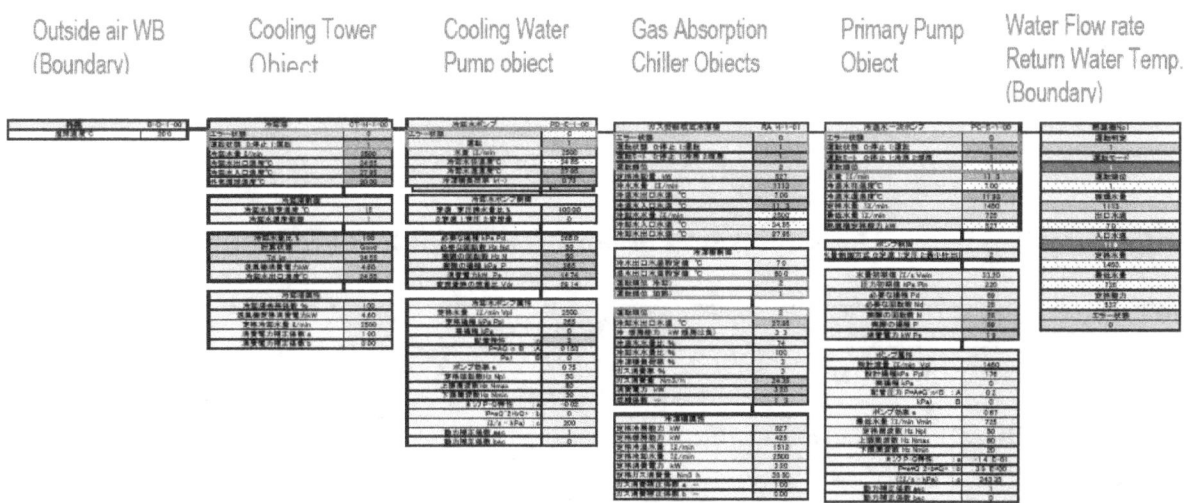

Figure 0-35 Example of system Configuration

Analytical Engine

Optimization of HVAC operation is simulated by convergent calculation using ad-in solvers in spreadsheet

End Users

The target end users are :

*   Designers;
*   Builders;
*   Building maintenance managers; and
*   Government officials responsible for public building planning, operations and maintenance.

The benefits of this tool are that it :

- Performs management taking into account operational status in off-peak period;
- Quantitatively supports plan, design, and operation;
- Monitors continuous performance assurance using consistent indicators throughout lifecycle, and performs PDCA cycle;
- Detects and analyzes potential missed defects; and
- Securely performs operator training.

Target Systems to be tested

HVAC system including advanced HVAC components such as double/triple effect absorption heater-chiller in cogeneration system, cooling control with outdoor air, and thermal energy storage.

No. 15   SUBTASK B: Commissioning and Optimization of Existing Buildings

TOOL NAME	EnergyPlus+VRF
COUNTRY / ORGANIZATION	Japan / DAIKIN AIR-CONDITIONING AND ENVIRONMENTAL LAB
TARGETED BUILDING / SYSTEM TYPE(S)	VRF package systems for office buildings
CONTACT PERSON:	Sumio Shiochi

Objectives

This tool specifically analyzes the energy consumption in the building's variable refrigerant flow (VRF) air-conditioning system. The VRF system features multi indoor unit connection.

Functions

EnergyPlus simulates the energy consumption, cooling (heating) capacity, and thermal condition of each zone of a centralized A/C system.

This module has the heat-pump VRF (cooling/heating) and heat recovery VRF function. In the heat-pump VRF function, cooling/heating power consumption is simulated for each VRF outdoor unit, depending on the operating mode. In the heat recovery VRF function, power consumption is calculated by weight condition factors (cooling or heating).

In this module, energy consumption comparisons between VRF A/C systems and centralized A/C systems are available.

Data Management

The performance curve is used to calculate VRF power consumption. The performance curve is defined by the second order equation. These are the capacity characteristic, power input characteristic, partial load characteristic, and combination ratio characteristic (indoor-unit to outdoor-unit), etc. The calculation result is expressed as output according to EnergyPlus.

Implementation

This module is developed on the EnergyPlus platform. The VRF module is customized to the existing EnergyPlusUnitary module.

Operability

The VRF module's operability depends on the EnergyPlus IDF editor. All required VRF module input parameters are set in the IDF editor.

Analytical Engine

The calculation engine is the EnergyPlus solver. The VRF module is built into the EnergyPlus Solver.

End Users

The end user is a professional engineer licensed to evaluate HVAC system energy consumption. The simulation is used to optimize the building's air-conditioning system.

Benefit

By selecting high performance and energy saving air-conditioning systems, building owners in the US may receive a tax incentive, thereby reducing construction costs.

Target Systems to be Tested

There is no current plan.

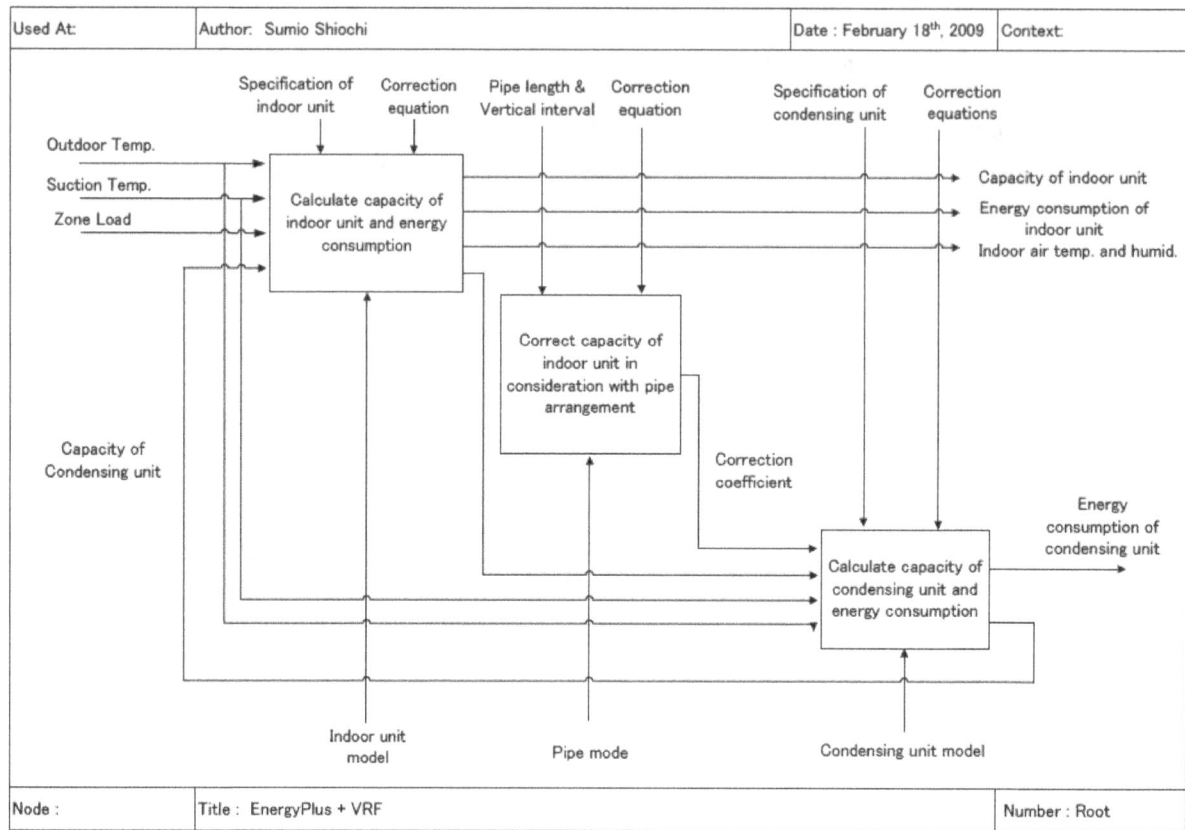

Specification of indoor unit Correction equation Pipe length & Vertical interval Correction equation Specification of condensing unit Correction equations

Outdoor Temp.
Suction Temp.
Zone Load

Calculate capacity of indoor unit and energy consumption

Capacity of indoor unit
Energy consumption of indoor unit
Indoor air temp. and humid.

Correct capacity of indoor unit in consideration with pipe arrangement

Capacity of Condensing unit

Correction coefficient

Calculate capacity of condensing unit and energy consumption

Energy consumption of condensing unit

Indoor unit model

Pipe mode

Condensing unit model

Node : | Title : EnergyPlus + VRF | Number : Root

Figure 0-36 IDEF representation of Energy Plus _VRF

TOOL NAME	CFD Coupled Simulation Tool for Natural Ventilation System with Earth Tube
COUNTRY / ORGANIZATION	Japan/SANKO Air Conditioning CO.,LTD.
TARGETED BUILDING / SYSTEM TYPE(S)	Natural Ventilation System with earth Tube
CONTACT PERSON:	Mingjie Zheng, Song Pan

Objectives

In the case of large spaces with exceptional vertical air temperature distribution, the simple method for natural ventilation systems cannot correctly represent the physical phenomenon, resulting in possible design fault at the planning phase. A CFD coupled simulation tool uses an earth tube to assess a natural ventilation system's air volume.

Functions

Figure 0-37 shows the IDEFØ diagram and tool functions. Figure 0-38 is a detailed diagram of Activity B0 from Figure 0-37.

The tool's main functions are:

- To use the earth tube to calculate the outlet air temperature. (Activity A0 in Figure 0-37)

- To calculate each aperture's air volume and flow direction. (Activity B0 in Figure 0-37)

- To calculate the room's vertical air temperature. (Activity C0 in Figure 0-37)

- To assess if the air volume at each aperture and the vertical air temperature become convergent. (Activity D0 and E0 in Figure 0-37)

Using this tool, we can predict the air volume of a natural ventilation system with an earth tube at the design phase. In the acceptance and operation phases, the model can be modified, to accurately predict the annual natural ventilation air volume and determine a system operation strategy.

Data Management

Input data:

- Weather data (wind velocity, wind direction, outdoor air temperature, solar radiation etc.);

- Earth tube parameters (size, length, depth, inlet air temperature, soil temperature, etc.);

- Parameters of natural ventilation apertures (number, height, area, etc.); and

- CFD parameters (room shape, boundary conditions, energy and loss coefficient of turbulent flow at supply air outlet, solar radiation, internal heat gain, etc.)

Output data:

- Air volume and flow direction of each aperture; and

- Vertical air temperature distribution.

Implementation

This tool uses CFD software available commercially, a natural ventilation analysis tool and an earth-air exchanger calculator developed by the authors.

Operability

CFD software is equipped with a convenient user interface. User interfaces for the natural ventilation analysis tool is in development.

Analytical Engine

CFD, natural ventilation analysis tool

End Users

Commissioning authority, System designer

Benefit

This tool is practical for assessing designed air volumes of natural ventilation systems with an earth tube.

Target Systems to be Tested

The tool was tested in a case study at an elementary school gymnasium in Toyama Province, Japan.

Used As:

Author: Mingjie Zheng and Song Pan (Sanko Air Conditioning CO.,LTD.)

Date: March 6th, 2009

Context

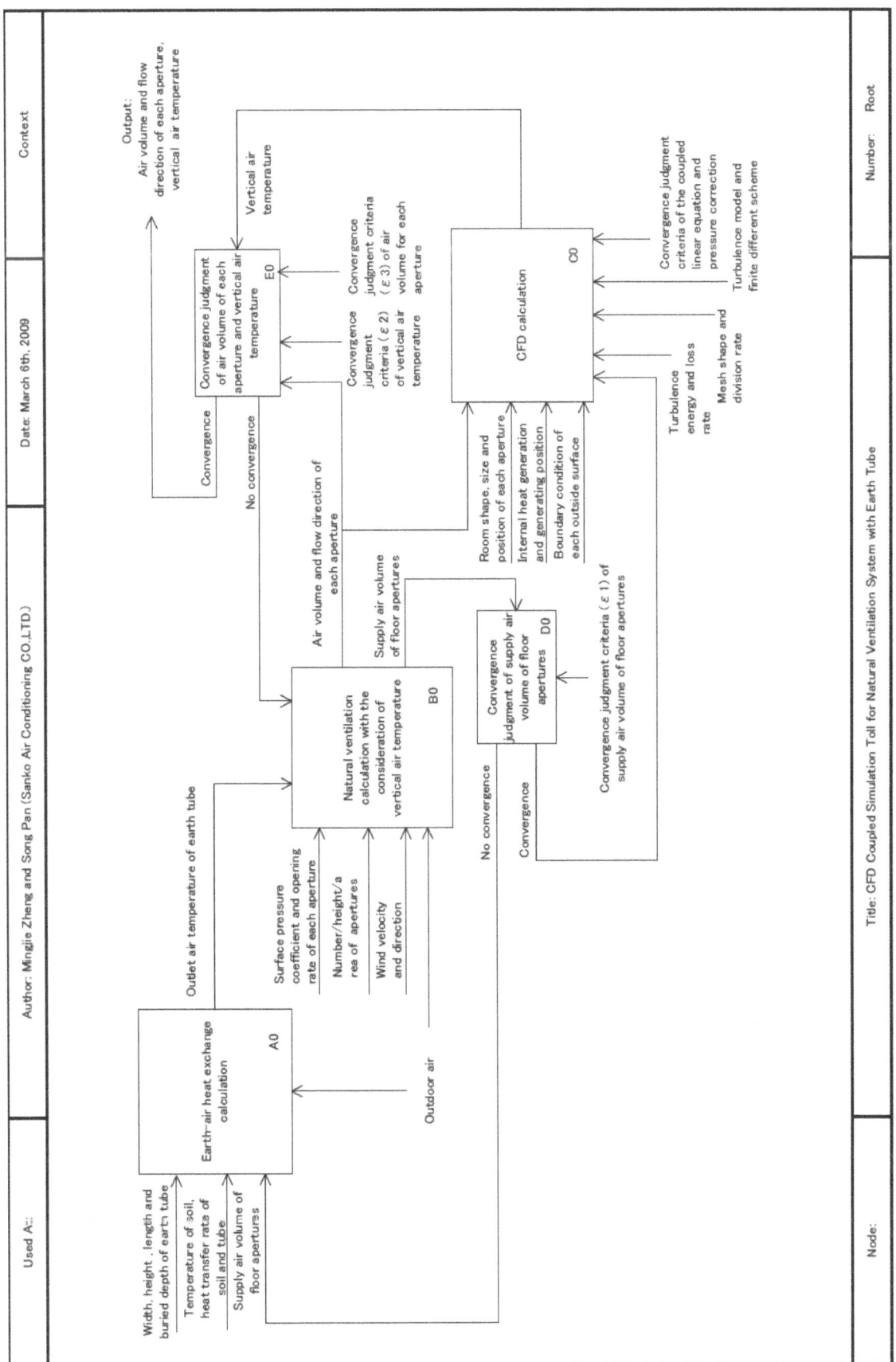

Figure 0-37 IDEF0 tool diagram (Root)

Used At:

Author: Mingjie Zheng and Song Pan (Sanko Air Conditioning CO.,LTD.)

Date: March 6th, 2009

Context

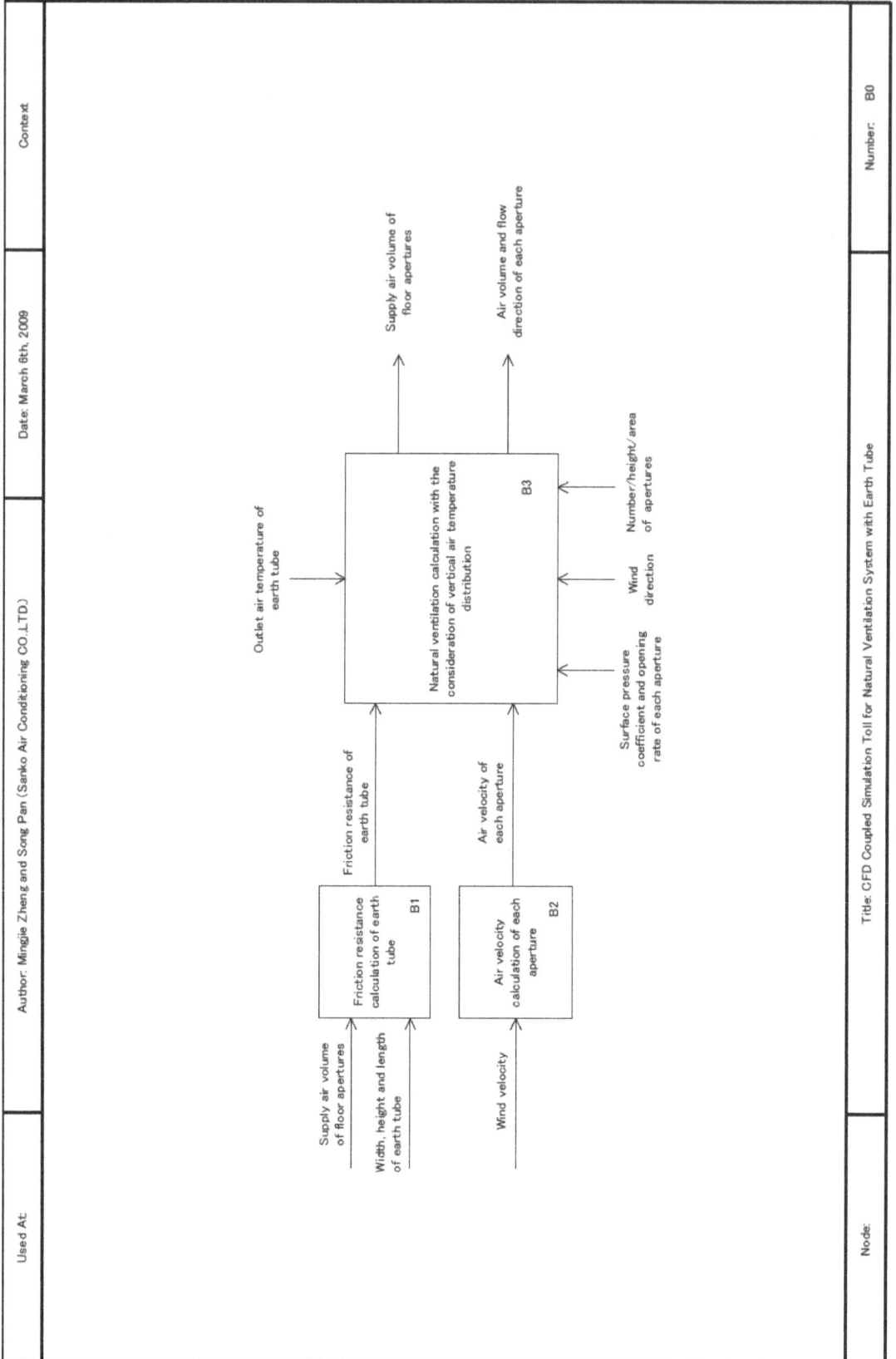

Figure 0-38 IDEFO tool diagram (B0)

TOOL NAME	FDD Proto-type tool for laboratory exhaust system
COUNTRY / ORGANIZATION	Japan
TARGETED BUILDING / SYSTEM TYPE(S)	Laboratory, factory
CONTACT PERSON:	Takao Odajima

Objectives

This tool's primary aim is the optimization of HVAC energy consumption in a laboratory, where many spot exhaust equipment units are installed. Spot exhaust equipment units are mandatory to maintain chemical and biological safety for laboratory employees. Optimization is achieved by enhancing exhaust flow rate. The tool's objective is to optimize supply and exhaust flow rates to save energy.

Functions

Figure 0-39 shows the IDEFØ tool diagram. Figure 0-40 is an expanded diagram of Activity A

The main function of the tool is:

- To grasp exhaust equipments' operation status, such as fume hoods.

- To grasp sash opening height of fume hoods (Activity A in Figure 0-39).

- To calculate exhaust flow rates of each fume hood (Activity B in Figure 0-39).

- To calculate exhaust flow rate of whole exhaust system.

- To calculate air supply flow rate for each room and whole supply system. (Activity C in Figure 0-39)

- To calculate energy consumption. (Activity D in Figure 0-39)

The details of Activity B are described in Figure 0-40, which are the expanded IDEFØ diagrams.

Data Management

Input data:

- Outdoor air condition/room temperature,
- Status of fume hoods (ON-OFF), opening height of fume hood,
- Flow rate of fresh air intake

Output data:

- Characteristics of fume hood operation
- Visual information of outstanding operation
- Action list

Implementation

The tool is developed without market software.

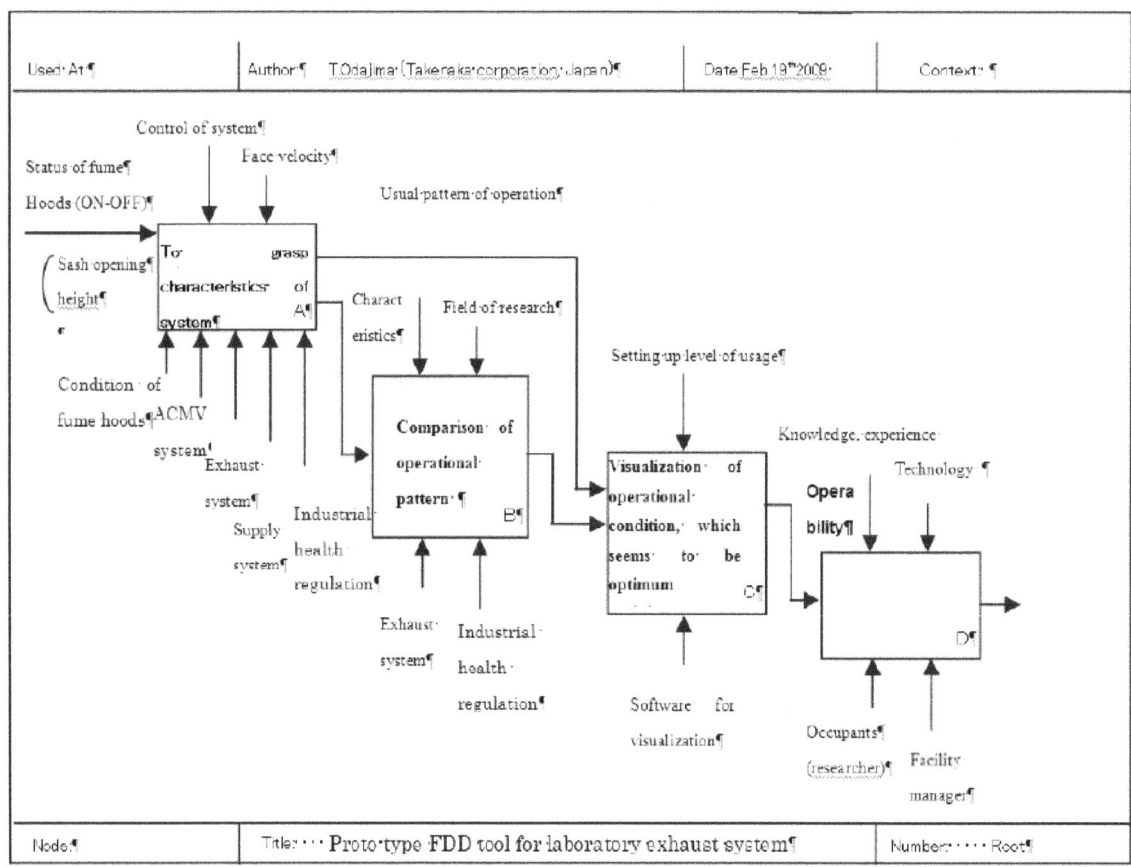

Figure 0-39 Root level IDEF representation

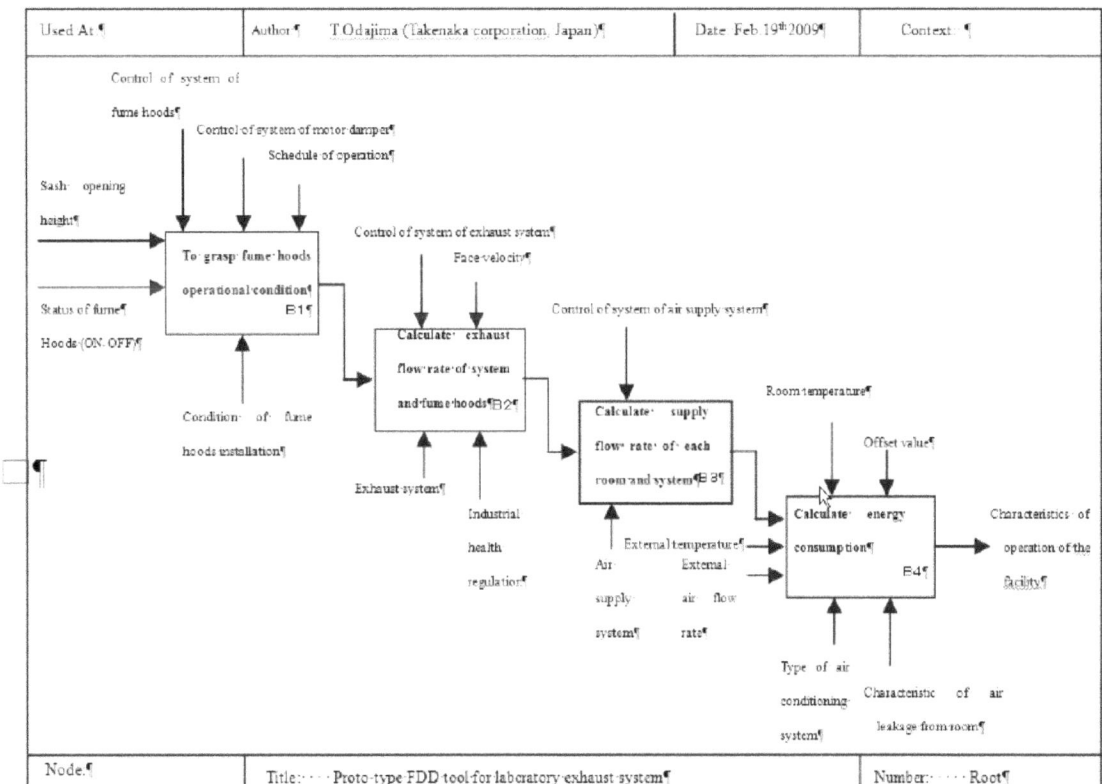

Used At ¶	Author ¶	T Odajima (Takenaka corporation, Japan)¶	Date Feb 19th 2009¶	Context ¶

Control of system of fume hoods¶

Control of system of motor damper¶

Schedule of operation¶

Sash opening height¶

Control of system of exhaust system¶

Face velocity¶

Control of system of air supply system¶

Room temperature¶

Offset value¶

Status of fume Hoods (ON-OFF)¶

To grasp fume hoods operational condition¶ B1¶

Calculate exhaust flow rate of system and fume hoods¶ B2¶

Calculate supply flow rate of each room and system¶ B3¶

Calculate energy consumption¶ B4¶

Characteristics of operation of the facility¶

Condition of fume hoods installation¶

Exhaust system¶

Industrial health regulation¶

External temperature¶

Air supply system¶

External air flow rate¶

Type of air conditioning system¶

Characteristic of air leakage from room¶

Node ¶	Title: Proto-type FDD tool for laboratory exhaust system¶	Number: Root¶

Figure 0-40 Root level IDEF representation (2)

Interface

Graphical interface is developed.

Analytical Engine

The tool is developed without market software .

End Users

Building operator, facility manager, laboratory occupants (researchers)

Benefit

The benefits to users are that:

- It may help facility managers detect an exhaust system fault, or the inappropriate behavior of occupants.

- It may call laboratory occupants' attention to take proper actions to minimize energy.

225

- It may optimize total energy consumption while maintaining chemical/biological safety.

Target Systems to be Tested

Tool is under development, and testing is planned at a pharmaceutical laboratory, upon completion in early 2011.

6.5 APPENDIX-5 Information Used in EB and LEB Cx tools Developed within Annex 47

This section provides the matrix of information used in developing each of the tools listed. Although the printed version of this document has limitations on viewing the entire window of embedded files, the full matrix can be obtained from the individual tool developers.

SUBTASK B: Commissioning and Optimization of Existing Buildings

TOOL NAME	DABO (for AHU)
COUNTRY/ORGANIZATION	Canada
TARGETED BUILDING/SYSTEM TYPE(S)	Commercial and Institutional
CONTACT PERSON:	Daniel Choinière

REQUIRED INFORMATION

	I) Design data (ex: area, component performance, etc)	II) Operation data (ex: set point, operation time, etc)	III) Measured data (ex: temperature, humidity, etc)
a) Building data (ex: area size, U values, solar shading, etc)			
b) Room environmental requirement or criteria (ex: room temperature set point, air quality, etc)	Return air humidity set point reasonable value Maximum reasonable value for return air humidity in the summer Minimum reasonable value for return air humidity in the summer Reasonable value for maximum return air temperature summer Reasonable value for rm	Zone temperature performance indice, First level alarm, Zone temperature performance indiceSecond level alarm, Zone Temperature performance indiceFirst level alarm, Zone pressure performance indiceSecond level alarm, Zone pressure performance indiceN	ControlZone/CO2Sensor/Status ControlZone/HumiditySensor/SetPoint ControlZone/HumiditySensor/Status ControlZone/PressureSensor/SetPoint ControlZone/PressureSensor/Status ControlZone/TemperatureSensor/SetPoint ControlZone/TemperatureSensor/Status
c) Weather data (ex: temperature, solar radiation, etc)	Design summer outdoor air temperature Design winter outdoor air temperature Outdoor CO2 maximum Outdoor CO2 minimum	Outside air temperature reference for heating degree calculation Outside air temperature reference for cooling degree calculation Threshold for zone air temperature setpoint	OutdoorCondition/CO2Sensor/Status OutdoorCondition/Enthalpy/Status OutdoorCondition/HumiditySensor/Status OutdoorCondition/TemperatureSensor/Status
d) Energy data (ex: energy prices, statistics of energy usage, etc)			

e)	HVAC system data (ex system type, configuration of system, etc)	System Type System configuration (location of sensors) Outside air available for warm duct(dual duct)	Differential design supply air temperature, cooling Differential design supply air temperature, heating Outside air available for warm duct Temperature rise across return fan Temperature rise across supply fan Recuperation coil influence on AHU cooling
f)	HVAC component data (ex COP, water flow rate, etc)	Pressure drop across filter – maximum Pressure drop across filter – minimum Supply / outdoor air humidity maximum value Supply / outdoor air humidity minimum value Outdoor air temperature maximum value Outdoor air temperature minimum value Supply/return CO2 maximum Supply/return CO2 minimum Return air humidity maximum value Return air humidity minimum value Return air temperature maximum value Return air temperature minimum value Recuperation coil influence on AHU heating Return fan maximum current Return fan minimum current Return fan maximum flow Return fan minimum flow Supply air pressure set point maximum Supply air pressure set point minimum Supply air temperature minimum value Supply air temperature maximum value Supply fan maximum current Supply fan minimum current Supply fan maximum flow Supply fan minimum flow Maximum limit value of source air temperature Minimum limit value of source air temperature Threshold for mixed air damper control and status Threshold for return / supply / mixed air temperature with cooling pump ON Threshold for return / supply / mixed air temperature with heating pump ON Threshold for Return / supply / mixed air temperature with no heating, no cooling or diagnosis mode Threshold for coil control and status Threshold, CO2 setpoint	Pressure drop across filter – maximum Threshold for mixed air damper control and status Threshold for return / supply / mixed air temperature with cooling pump ON Threshold for return / supply / mixed air temperature with heating pump ON Threshold for Return / supply / mixed air temperature with no cooling or diagnosis mode Threshold for coil control and status Threshold, CO2 setpoint Threshold for outside air flow Threshold for outside air humidity Threshold for outside air temperature Threshold for Return / supply air humidity Threshold for reasonable value for return air temperature Threshold supply / return air flow Threshold on enthalpy

229

Control system data g) (ex set point, PID parameters, etc)	Winter night freeze protection shut off, outside air temperature	Outside air temperature switch over mode. free cooling to mechanical cooling maximum
	Mixed air damper at minimum position	Outside air temperature switch over mode. free cooling to mechanical cooling minimum
	Maximum reasonable mixed air temperature in Off mode.	Maximum outside air temperature for heating switch over mode
	Minimum reasonable mixed air temperature in Off mode.	Outside air temperature switch over mode. heating to free cooling maximum
	Percentage, minimum air flow	Outside air temperature switch over mode. heating to free cooling minimum
	Supply air temperature reasonable maximum value for free cooling and cooling mode	Outside air temperature switch over mode. heating to free cooling threshold
	Supply air temperature reasonable minimum value for free cooling and cooling mode	Limit outside air temperature to satisfy a zone air temperature without cooling device.
	Supply air temperature maximum value for heating mode	Fresh air damper at minimum position
	Supply air temperature minimum value for heating mode	Threshold for return air humidity control and status
	Supply/return differential airflow	Maximum threshold value in mixed air temperature threshold calculation
	Mixed air temperature maximum value	Minimum threshold value in mixed air temperature threshold calculation
	Mixed air temperature minimum value	Maximum (SOAT-RAT) in mixed air temperature threshold calculation
	Minimum fresh air flow	Threshold, preheat air temperature setpoint
		Threshold, percentage of fresh air
		Threshold for return air humidity set point
		Threshold Supply air pressure set point
		Threshold for supply air temperature setpoint
		Threshold, supply air temperature setpoint
		Threshold supply/return fan volume command/status

SUBTASK B: Commissioning and Optimization of Existing Buildings

TOOL NAME	CITE-AHU automated commissioning tool for
COUNTRY /	USA/NIST
TARGETED BUILDING / SYSTEM TYPE(S)	AHU
CONTACT PERSON:	Natascha Milesi

REQUIRED

	I) Design (ex. area, component performance, etc.)	II) Operation (ex. set point, operation time.)	III) Measured data (ex. temperature, humidity, etc.)
Building a) (ex. area size, U values, solar shading, etc.)			
Room environmental criteria b) (ex. room temperature set, air quality.)			Room air temperature and humidity
Weather data c) (ex. temperature, solar etc.)			Outdoor air temperature and humidity
Energy data d) (ex. energy prices, statistics energy usage, etc)			
HVAC system e) (ex. system type, configuration, system, etc)	Schematic diagram of system, system type, structure linking building sub zones, and data	Operation	
HVAC component f) (ex. COP, water flow rate, etc.)	Specification of component, i.e. defrost recovery, humidification column		
Control system g) (ex. set point, PID parameters, etc)	Control strategy of VAV control and control	Set point of water flow rate to the ground storage period ( ), inverter value of cooling tower fan (30 %), and outlet temperature of cooling coil (°C)	-Supply return, mixed, and outdoor temperature. Outdoor air humidity and/or air humidity and/or enthalpy (for based economizer control). Heating coil signal, cooling coil valve signal, damper signal
Maintenance h) (ex. maintenance level, number of maintainer, etc)			sensor

231

SUBTASK B: Commissioning and Optimization of Existing Buildings

TOOL NAME	A FDD tool "I-BIG" for AHU's.
COUNTRY / ORGANIZATION	The Netherlands / TNO Built Environment and Geosciences
TARGETED BUILDING / SYSTEM TYPE(S)	In general AHU's.
CONTACT PERSON:	Luc Soethout, Henk Peitsman
REQUIRED INFORMATION	

	I) Design data (ex. area, component performance, etc.)	II) Operation data (ex. set point, operation time, etc.)	III) Measured data (ex. temperature, humidity, etc.)
a) Building data (ex. area size, U values, solar shading, etc.)			
b) Room environmental requirement or criteria (ex. room temperature set point, air quality, etc.)		Room air temperature setpoint, room air relative humidity setpoint	Room air temperature, room air relative humidity
c) Weather data (ex. temperature, solar radiation, etc.)			Outdoor air temperature and humidity
d) Energy data (ex. energy prices, statistics of energy usage, etc.)			
e) HVAC system data (ex. system type, configuration of system, etc.)	Schematic diagram of AHU system	weekly operation schedule	ON-OFF signal AHU's
f) HVAC component data (ex. COP, water flow rate, etc.)	Specification of constant airflow measure rings, maximum airflow rate of AHU's		Flow rate supply- and return fan, temperature in duct after outlet solar collector, outlet temperature solar collector, supply air temperature, supply air temperature setpoint, temperature after heat-exchanger, inlet temperature solar collector
g) Control system data (ex. set point, PID parameters, etc.)	Control strategy of AHU, max frequency value of frequency controllers fan's		Setpoints, Control signals (0-100%),
h) Maintenance data (ex. maintenance level, mumble of maintainer, etc.)			

SUBTASK B: Commissioning and Optimization of Existing Buildings

TOOL NAME	Performance Analysis Tool for Heating System
COUNTRY / ORGANIZATION	Norwegina University of Science and Technology
TARGETED BUILDING / SYSTEM TYPE(S)	Hydronic Heating System
CONTACT PERSON:	Natasa Djuric

REQUIRED INFORMATION

	I) Design data (ex. area, component performance, etc.)	II) Operation data (ex. set point, operation time, etc.)	III) Measured data (ex. temperature, humidity, etc.)
a) Building data (ex. area size, U values, solar shading, etc.)	Area size: 13700 m² Avreage U vaule for zones/rooms A general building plan Building zones		
b) Room environmental requirement or criteria (ex. room temperature set point, air quality, etc.)		Indoor air temperature	Indoor air temperature
c) Weather data (ex. temperature, solar radiation, etc.)		Outdoor temperature	Outdoor temperature
d) Energy data (ex. energy prices, statistics of energy usage, etc.)		Hourly energy consumption of the system	- The hot water energy consumption - The energy consumption of branches that do not have measurements in the control system
e) HVAC system data (ex. system type, configuration of system, etc.)	Schematic diagram of system		
f) HVAC component data (ex. COP, water flow rate, etc.)	Specification of components: The size/capacity of radiators in each room, The capacity of the heat exchangers in the substation, The valves characteristics in the substation		
g) Control system data (ex. set point, PID parameters, etc.)	Design supply temperature, Design temperature difference	- Heating supply temperature curve (specification of the important points) - Operational schedule	- Heating supply temerature - Hot water supply temperature
h) Maintenance data (ex. maintenance level, mumble of maintainer, etc.)		1. Maintenance service hired in building 2. Control system maintenance: technical support from the control system producer	

TOOL NAME	A tool to be used for energy analysis and fault detection on the whole building level		
COUNTRY / ORGANIZATION	Germany / Fraunhofer ISE		
TARGETED BUILDING / SYSTEM TYPE(S)	non-residential (office) buildings / overall performance		
CONTACT PERSON:	Christian Neumann		
REQUIRED INFORMATION			
	I) Design data (ex area, component performance, etc)	II) Operation data (ex set point, operation time, etc)	III) Measured data (ex temperature, humidity, etc)
a) Building data (ex area size, U values, solar shading, etc)	net floor area, utilization of zones, information about internal aims (size, distribution)		
b) Room environmental requirement or criteria (ex room temperature set point, air quality, etc)		requirements for indoor climate depending on the kind of system	temperature, moisture of selected reference zones
c) Weather data (ex temperature, solar radiation, etc)			temperature , moisture, solar radiation (global horizontal)
d) Energy data (ex energy prices, statistics of energy usage, etc)	if available: historic consumption data and/or planned energy demand		total delivered energy (fuel, district heat/cold, electricity) AND WATER CONSUMPTION
e) HVAC system data (ex system type, configuration of system, etc)	kind of system, schemes, zonal distribution	operation schedules	Supply and return temperatures of major distribution (water and air), in case of air: moisture of exhaust and supply if treated
f) HVAC component data (ex COP, water flow rate, etc)	if available: flow rates, capacities	operation schedules	
g) Control system data (ex set point, PID parameters, etc)	if available: control strategy of AHUs, pumps, different operation modes		control signals of drives (pumps, fans)
h) Maintenance data (ex maintenance level, mumble of maintainer, etc)			

SUBTASK B: Commissioning and Optimization of Existing Buildings

TOOL NAME	Optimization tool for air-conditioning system operation considering thermal load prediction errors
COUNTRY / ORGANIZATION	Japan
TARGETED BUILDING / SYSTEM TYPE(S)	Exsisted office buildings and centered HVAC systems
CONTACT PERSON:	Yasunori Akashi, Daisuke Sumiyoshi

REQUIRED INFORMATION

	I) Design data (ex. area, component performance, etc.)	II) Operation data (ex. set point, operation time, etc.)	III) Measured data (ex. temperature, humidity, etc.)
a) Building data (ex. area size, U values, solar shading, etc.)	General information for calculating building thermal loads such as floor plans, sections, windows size, thermal properties of walls/windows, etc.	Occupancy time	
b) Room environmental requirement or criteria (ex. room temperature set point, air quality, etc.)	Room temperature set point		Indoor air temperature and humidity
c) Weather data (ex. temperature, solar radiation, etc.)			Outdoor air temperature and humidity, solar radiation
d) Energy data (ex. energy prices, statistics of energy usage, etc.)			System and components energy consumption
e) HVAC system data (ex. system type, configuration of system, etc.)	Schematic diagram of system	Operation time	System energy consumption and amount of heat dealt with the system
f) HVAC component data (ex. COP water flow rate, etc.)	Specification of components	Operation time	Components energy consumption and amount of heat dealt with the components, physical state quantities such as water temperature, flow rate, pressure, etc.
g) Control system data (ex. set point, PID parameters, etc.)	System control logics and strategies including control set points, PID parameters, etc.		
h) Maintenance data (ex. maintenance level, mumble of maintainer, etc.)			

SUBTASK B: Commissioning and Optimization of Existing Buildings

TOOL NAME	An on-going commissioning tool for VRV package systems
COUNTRY / ORGANIZATION	Japan / Chubu University
TARGETED BUILDING / SYSTEM TYPE(S)	All types / VRV package systems
CONTACT PERSON:	Motoi Yamaha

REQUIRED INFORMATION

	I) Design data (ex. area, component performance, etc.)	II) Operation data (ex. set point, operation time, etc.)	III) Measured data (ex. temperature, humidity, etc.)
a) Building data (ex. area size, U values, solar shading, etc.)	area, U values, solar shading.... data required for heat load calculations		
b) Room environmental requirement or criteria (ex. room temperature set point, air quality, etc.)			room temperature and room humidity
c) Weather data (ex. temperature, solar radiation, etc.)	Standard weather data for heat load calculation		temperature, humidity, insolation
d) Energy data (ex. energy prices, statistics of energy usage, etc.)			power consumption
e) HVAC system data (ex. system type, configuration of system, etc.)		Only adaptable for VRV package systems	
f) HVAC component data (ex. COP, water flow rate, etc.)	Nominal capacity, nominal power consumption, number of compressors, type of refrigerant, nominal COP		power consumption
g) Control system data (ex. set point, PID parameters, etc.)			
h) Maintenance data (ex. maintenance level, mumble of maintainer, etc.)			

236

SUBTASK B: Commissioning and Optimization of Existing Buildings

TOOL NAME	A tool to monitor continuously oprational data of each HVAC equipment
COUNTRY / ORGANIZATION	Hitachi Plant / Japan
TARGETED BUILDING / SYSTEM TYPE(S)	Office buildings
CONTACT PERSON:	Hiroo Sakai
REQUIRED INFORMATION	

	I) Design data (ex. area, component performance, etc.)	II) Operation data (ex. set point, operation time, etc.)	III) Measured data (ex. temperature, humidity, etc.)
a) Building data (ex. area size, U values, solar shading, etc.)			
b) Room environmental requirement or criteria (ex. room temperature set point, air quality, etc.)		Set points for room temperature, humidity control	Room Tempeature Relative Humidity
c) Weather data (ex. temperature, solar radiation, etc.)			Dry bulb temperature Wet bulb temperature Relative Humidity
d) Energy data (ex. energy prices, statistics of energy usage, etc.)			
e) HVAC system data (ex. system type, configuration of system, etc.)	Configuration of the system	Supply/ returen temperature of chilled water and condenser water Flow rate of chilled water, condenser water Temperature of supply air	Supply/ returen temperature of chilled water and condenser water Flow rate of chilled water, condenser water Temperature of supply air
f) HVAC component data (ex. COP, water flow rate, etc.)	Design data of each component Allowable operational condition		
g) Control system data (ex. set point, PID parameters, etc.)	Design set points for each component		
h) Maintenance data (ex. maintenance level, mumble of maintainer, etc.)			

237

SUBTASK B: Commissioning and Optimization of Existing Buildings

TOOL NAME	Initial Cx Tool for HVAC System in Large Enclosure
COUNTRY / ORGANIZATION	Japan /SANKO Air Conditioning CO.,LTD
TARGETED BUILDING / SYSTEM TYPE(S)	Indoor climate analysis for the large enclosures
CONTACT PERSON:	Mingjie Zheng

REQUIRED INFORMATION

	I) Design data (ex. area, component performance, etc.)	II) Operation data (ex. set point, operation time, etc.)	III) Measured data (ex. temperature, humidity, etc.)
a) Building data (ex. area size, U values, solar shading, etc.)	Space size, Window size, Over-all heat transfer coefficient, material characteristics of building skin	Warming up time, Size and shape of separated mesh for CFD	Outside surface temperatures of building skin
b) Room environmental requirement or criteria (ex. room temperature set point, air quality, etc.)	Room temperature set point, Pick-up time	Model of turbulent flow calculate, Finite difference scheme	Indoor air temperature gradient
c) Weather data (ex. temperature, solar radiation, etc.)	Outdoor air temperature, Solar radiation, Wind speed		Outdoor air temperature, Solar radiation, Wind speed
d) Energy data (ex. energy prices, statistics of energy usage, etc.)	Size and position of lighting, furniture,office-machine and occupancy	Energy and loss coefficient of turbulent flow in supply air outlet	Heat gain and schedule from lighting, office-machine and occupancy
e) HVAC system data (ex. system type, configuration of system, etc.)	Schematic diagram of system	Operation time	Operation time
f) HVAC component data (ex. COP, water flow rate, etc.)	Size and shape of supply air outlet and return air inlet	Supply air volume,Temperature/humidity of supply air	Air temperature in duct and pipe
g) Control system data (ex. set point, PID parameters, etc.)	Room temperature set point, PID parameters of controller for the supply air temperature, Time constant and sensitivity of thermostat	Integral sampling step, Control step of AHU water valve opening, Time delay, Dead band of roomair temperature sensor	Supply air volume,Temperature/humidity of supply air, AHU water valve opening
h) Maintenance data (ex. maintenance level, mumble of maintainer, etc.)			

238

SUBTASK B: Commissioning and Optimization of Existing Buildings

TOOL NAME	On-going Type FDD Tool with Pattern Recognition for VAV Systems
COUNTRY / ORGANIZATION	Japan /SANKO Air Conditioning CO.,LTD
TARGETED BUILDING / SYSTEM TYPE(S)	non-residential (office) buildings / performance
CONTACT PERSON:	Mingjie Zheng and Song Pan

REQUIRED INFORMATION

	I) Design data (ex. area, component performance, etc.)	II) Operation data (ex. set point, operation time, etc.)	III) Measured data (ex. temperature, humidity, etc.)
a) Building data (ex. area size, U values, solar shading, etc.)			
b) Room environmental requirement or criteria (ex. room temperature set point, air quality, etc.)		Set point of room air temperature	Air temperature of each room
c) Weather data (ex. temperature, solar radiation, etc.)			
d) Energy data (ex. energy prices, statistics of energy usage, etc.)			
e) HVAC system data (ex. system type, configuration of system, etc.)	Number of VAV units	Sampling intervals of system's data, Threshold of statistical data of measured data and their Fourier Transform values	Operation state of HVAC system, Supply air temperature and volume of AHU, Air volume of VAV unit,
f) HVAC component data (ex. COP, water flow rate, etc.)		Sampling intervals of unit's data	Inlet/outlet cooling water temperature of AHU, Cooling water flow rate of AHU
g) Control system data (ex. set point, PID parameters, etc.)			
h) Maintenance data (ex. maintenance level, mumble of maintainer, etc.)			

TOOL NAME	A Tool for optimizing the operation of heating cooling plants
COUNTRY / ORGANIZATION	Japan / Kyoto University
TARGETED BUILDING / SYSTEM TYPE(S)	Any type
CONTACT PERSON:	Harunori Yoshida

REQUIRED INFORMATION

	I) Design data (ex. area, component performance, etc.)	II) Operation data (ex. set point, operation time, etc.)	III) Measured data (ex. temperature, humidity, etc.)
a) Building data (ex. area size, U values, solar shading, etc.)	Hourly heating/cooling load		Hourly heating/cooling load
b) Room environmental requirement or criteria (ex. room temperature set point, air quality, etc.)			
c) Weather data (ex. temperature, solar radiation, etc.)	Hourly temperature, relative humidity		Hourly temperature, relative humidity
d) Energy data (ex. energy prices, statistics of energy usage, etc.)	Energy prices		Hourly energy cost of each heating/cooling source equipments (for the purpose of refining equipment models)
e) HVAC system data (ex. system type, configuration of system, etc.)	Specifications of heating/cooling source equipments, delivery pumps, cooling towers, thermal storage tanks, heat exchangers used for thermal storage systems		
f) HVAC component data (ex. COP, water flow rate, etc.)		Set points of chilled/hot water temperature, cooling water temperature	
g) Control system data (ex. set point, PID parameters, etc.)	Control logic for start/stop heating/cooling source equipments, cooling tower fans		
h) Maintenance data (ex. maintenance level, mumble of maintainer, etc.)			

SUBTASK B: Commissioning and Optimization of Existing Buildings

TOOL NAME	A Tool for Optimizing the HVAC System with Ground Thermal Storage
COUNTRY / ORGANIZATION	Japan / Kyoto University
TARGETED BUILDING / SYSTEM TYPE(S)	HVAC System with Underground Thermal Storage
CONTACT PERSON:	Harunori Yoshida

REQUIRED INFORMATION

	I) Design data (ex area, component performance, etc)	II) Operation data (ex set point, operation time, etc)	III) Measured data (ex temperature, humidity, etc)
a) Building data (ex area size, U values, solar shading, etc)	Site/Area size Foundation pile/Number Foundation pile/Diameter Foundation pile/Depth Water pipe/Shape Water pipe/length Water pipe/Inside diameter Water pipe/Outside diameter Soil/Heat thermal conductivity Soil/Specific Heat Soil/Density Concrete/Heat the		
b) Room environmental requirement or criteria (ex room temperature set point, air quality, etc)			Return air condition/Temperature Return air condition/Humidity
c) Weather data (ex temperature, solar radiation, etc)			Outdoor condition/Temperature Outdoor condition/Humidity
d) Energy data (ex energy prices, statistics of energy usage, etc)			DHC plant/System COP
e) HVAC system data (ex system type, configuration of system, etc)	HVAC system/System Type HVAC system/Configuration	HVAC system/Mode (Storage or discharge) HVAC system/Switch	
f) HVAC component data (ex COP, water flow rate, etc)	Pump for storage/Specification data Pump for discharge/Specification data Cooling Tower/Specification data Air heat exchanger/Specification data Cooling coil/Specification data		Pump for storage/Energy consumption Pump for discharge/Energy consumption Cooling Tower/fan/Energy consumption
g) Control system data (ex set point, PID parameters, etc)	Cooling coil/Water valve/Control strategy (the flow rate is controlled so that the outlet water temperature of the coil attains the temperature set point)	Pump for storage/water flow rate/Set point Cooling Tower/Fan/Inverter value/Set point Cooling Coil/Outlet water temperature/Set point	Ground/Charged heat amount Ground/Discharged heat amount Ground/Soil temperatures
h) Maintenance data (ex maintenance level, mumble of maintainer, etc)			

241

SUBTASK B: Commissioning and Optimization of Existing Buildings

TOOL NAME	A Tool to Estimate Energy Baselines
COUNTRY / ORGANIZATION	Japan / Kyoto University
TARGETED BUILDING / SYSTEM TYPE(S)	Any type
CONTACT PERSON:	Harunori Yoshida

REQUIRED INFORMATION

	I) Design data (ex. area, component performance, etc.)	II) Operation data (ex. set point, operation time, etc.)	III) Measured data (ex. temperature, humidity, etc.)
Building data a) (ex. area size, U values, solar shading, etc.)	Area Size (6169 m²). Air-conditioned area (3942 m²). Wall configuration		
Room environmental requirement or criteria b) (ex. room temperature set point, air quality, etc.)		Number of Occupants (260 persons), Calorific value of lighting and the office machines	Room air temperature and humidity
Weather data c) (ex. temperature, solar radiation, etc.)			Outdoor air temperature and humidity, Grobal solar radiation
Energy data d) (ex. energy prices, statistics of energy usage, etc.)			Energy consumption of whole building
HVAC system data e) (ex. system type, configuration of system, etc.)		Operating time	Fresh air intake volume (2.46 m³/hm²)
HVAC component data f) (ex. COP, water flow rate, etc.)			Inlet and outlet water temperature of chiller, water flow rate of chiller
Control system data g) (ex. set point, PID parameters, etc.)			
Maintenance data h) (ex. maintenance level, mumble of maintainer, etc.)			

SUBTASK B: Commissioning and Optimization of Existing Buildings

TOOL NAME	Tools for Life Cycle Energy Management (LCEM) for Buildings
COUNTRY / ORGANIZATION	Japan / Ministry of Land, Infrastructure and Transport (MLIT)
TARGETED BUILDING / SYSTEM TYPE(S)	Public buildings
CONTACT PERSON:	Public Building Association of Japan

REQUIRED INFORMATION

	I) Design data (ex. area, component performance, etc.)	II) Operation data (ex. set point, operation time, etc.)	III) Measured data (ex. temperature, humidity, etc.)
a) Building data (ex. area size, U values, solar shading, etc.)	floor area location items for cooling/heating load estimation annual operation schedule		
b) Room environmental requirement or criteria (ex. room temperature set point, air quality, etc.)	hourly indoor cooling/heating load	set point of indoor temparature (DB,WB) on/off time	indoor temparature (DB,WB)
c) Weather data (ex. temperature, solar radiation, etc.)	standard annual weather data		outside air temparature(DB, WB) supply water temparature
d) Energy data (ex. energy prices, statistics of energy usage, etc.)	gas calolific value volum of wasete heat from CGS	gas calolific value amount of use of wasete heat from CGS	actual gas consumption actual power consumption actual water consumption
e) HVAC system data (ex. system type, configuration of system, etc.)	system type, capacity, ducts specification system COP at partial load	HVAC on-off time operation schedule	
f) HVAC component data (ex. COP, water flow rate, etc.)	specifications of components chiller/heater, pump, AHU, fan, cooling tower, etc. COP at partial load		COP at actual load
g) Control system data (ex. set point, PID parameters, etc.)	Set point of chilled water temparature for chillers cooling water for cooling towers required water head for pumps	Set point of min and/or max chilled water temparature for chillers cooling water for cooling towers required water head for pumps	actual data of inlet and/or outlet of chilled/hot water temparature cooling water temparature
h) Maintenance data (ex. maintenance level, mumble of maintainer, etc.)	frequency of maintenance by components COP recovery rate		

243

TOOL NAME	Energy analysis module (on Energy Plus) for the variable refrigerant flow(VRF) air-conditioning system
COUNTRY / ORGANIZATION	Japan / Daikin Air-conditioning and Environmental Laboratory., Ltd
TARGETED BUILDING / SYSTEM TYPE(S)	HVAC System for the variable refrigerant flow(VRF) air-conditioning system
CONTACT PERSON:	Shingo Itoh

SUBTASK B: Commissioning and Optimization of Existing Buildings

REQUIRED INFORMATION

	I)	II)	III)
	Design data (ex. area, conponent performance, etc.)	Operation data (ex. set point, operation time, etc.)	Measured data (ex. temperature, humidity, etc.)
a) Building data (ex. area size, U values, solar shading, etc.)	Location ,Area size Structure of building (Thickness,Material properties, Window-size etc) Zoning		
b) Room environmental requirement or criteria (ex. room temperature set point, air quality, etc.)	Set point Ventilation Infiltration	Internal load driving schedule(People, lighting, Electric equipment, etc)	
c) Weather data (ex. temperature, solar radiation, etc.)	Typical Meteorological Year 2 (TMY) weather format.		
d) Energy data (ex. energy prices, statistics of energy usage, etc.)			
e) HVAC system data (ex. system type, configuration of system, etc.)	System type Configuration of system	Operation data	
f) HVAC component data (ex. COP, water flow rate, etc.)	Performance curve(Total capacity,Power input,PLR,Combination ratio)		
g) Control system data (ex. set point, PID parameters, etc.)			
h) Maintenance data (ex. maintenance level, mumble of maintainer, etc.)			

244

SUBTASK B: Commissioning and Optimization of Existing Buildings

TOOL NAME	CFD Coupled Simulation Tool for Natural Ventilation System with Earth Tube
COUNTRY / ORGANIZATION	Japan /SANKO Air Conditioning CO.,LTD
TARGETED BUILDING / SYSTEM TYPE(S)	annexation system of cool/heat tube and natural ventilation
CONTACT PERSON:	Mingjie Zheng, Song Pan

REQUIRED INFORMATION

	I) Design data (ex. area, component performance, etc.)	II) Operation data (ex. set point, operation time, etc.)	III) Measured data (ex. temperature, humidity, etc.)
a) Building data (ex. area size, U values, solar shading, etc.)	Width,height,length and buried depth of earth tube	mesh shape and division rate for CFD, Energy and loss coefficient of turbulent flow	Outside surface temperatures of building skin, Temperature of soil
b) Room environmental requirement or criteria (ex. room temperature set point, air quality, etc.)		Model of turbulent flow calculate, Finite difference scheme, Temperature of soil, Heat transfer rate of soil and tube	Indoor air temperature gradient
c) Weather data (ex. temperature, solar radiation, etc.)	Outdoor air temperature, Solar radiation, Wind speed, Wind direction		Outdoor air temperature, Solar radiation, Wind speed, wind direction
d) Energy data (ex. energy prices, statistics of energy usage, etc.)	Size and position of lighting, furniture,office-machine and occupancy	Heat gain and schedule from lighting, office-machine and occupancy	
e) HVAC system data (ex. system type, configuration of system, etc.)	Surface pressure coefficient and operating rate of each aperture, Number/height/area of apertures		Supply air temperature and volume of each aperture
f) HVAC component data (ex. COP, water flow rate, etc.)			
g) Control system data (ex. set point, PID parameters, etc.)		Convergence judgement criteria of the coupled linear equation and pressure correction	
h) Maintenance data (ex. maintenance level, mumble of maintainer, etc.)			

SUBTASK B: Commissioning and Optimization of Existing Buildings

TOOL NAME	Proto-type FDD tool for laboratory exhaust system
COUNTRY / ORGANIZATION	Japan / TAKENAKA CORPORATION
TARGETED BUILDING / SYSTEM TYPE(S)	Laboratory factory
CONTACT PERSON:	Takao Odajima

REQUIRED INFORMATION

	I) Design data (ex. area, component performance, etc.)	II) Operation data (ex. set point, operation time, etc.)	III) Measured data (ex. temperature, humidity, etc.)
a) Building data (ex. area size, U values, solar shading, etc.)			
b) Room environmental requirement or criteria (ex. room temperature set point, air quality, etc.)		Room temperature set point, Relative humidity set point, Room air pressure difference set point,	Room temperature , Relative humidity , Room air pressure difference
c) Weather data (ex. temperature, solar radiation, etc.)			Hourly temperature, relative humidity
d) Energy data (ex. energy prices, statistics of energy usage, etc.)	Energy prices		Hourly energy consumption of each heating/cooling source equipments
e) HVAC system data (ex. system type, configuration of system, etc.)	Schematic diagram of system, System type, Ahu zones	Operation time	Supply and exhaust air volume,
f) HVAC component data (ex. COP, water flow rate, etc.)	Specifications of HVAC equipments		Exhaust VAV status, Operation status of each fume hood operation, Sash opening height
g) Control system data (ex. set point, PID parameters, etc.)	Control strategy of supply and exhaust VAV control, supply and exhaust fan control and set point control	PID parameters	
h) Maintenance data (ex. maintenance level, mumble of maintainer, etc.)		Assumed demand factor based on past operational data	Demand factor of fume hood operation

SUBTASK B: Commissioning and Optimization of Existing Buildings

TOOL NAME	Energy Performance Commissioning of Existing Buildings and Building Portfolios	
COUNTRY / ORGANIZATION	Finland/VTT	
TARGETED BUILDING / SYSTEM TYPE(S)	All kind of existing buildings	
CONTACT PERSON:	Jorma Pietilainen	

REQUIRED INFORMATION

	I) Design data (ex. area, component performance, etc.)	II) Operation data (ex. set point, operation time, etc.)	III) Measured data (ex. temperature, humidity, etc.)
a) Building data (ex. area size, U values, solar shading, etc.)	Building type (purpose of use, classification based on the main use)		Minimum is the usable net floor area, possibly heated+total building volume
b) Room environmental requirement or criteria (ex. room temperature set point, air quality, etc.)	Share of spaces with "abnormal" condition requirements		
c) Weather data (ex. temperature, solar radiation, etc.)			Mean outdoor temperature and monthly/yearly heating (cooling) degree days, possibly other weather data as well
d) Energy data (ex. energy prices, statistics of energy usage, etc.)		Energy forms/carriers, Prices for heating and electrical energy+water	Yearly or (preferably) monthly consumption figures (preferably meeter readings) of heating and electrical energy +water
e) HVAC system data (ex. system type, configuration of system, etc.)		Possibly operation times or classification (8/12/16/24 hours/day)	System type for statistical analyses (classification like natural/mechanical exhaust air/mechanical intake+exhaust air/air conditning/heat recovery
f) HVAC component data (ex. COP, water flow rate, etc.)			
g) Control system data (ex. set point, PID parameters, etc.)		Control principles, manual or BAS or other automation/remote control (e.g. BEMS)	
h) Maintenance data (ex. maintenance level, mumble of maintainer, etc.)		M&O-manual (form paper/eelctronic)	
i) User data			Amount of users/year or month (if available) like pupils/students in schools etc. to be used in benchmarking and performance metrics